DEPARTMENT OF THE ARMY
U.S. Army Corps of Engineers
Washington, DC 20314-1000

EM 1110-2-1100
(Change 2)

CECW-CE

Manual
No. 1110-2-1100

1 April 2008

Engineering and Design
COASTAL ENGINEERING MANUAL

1. Purpose. The purpose of the *Coastal Engineering Manual* (CEM) is to provide a comprehensive technical coastal engineering document. It includes the basic principles of coastal processes, methods for computing coastal planning and design parameters, and guidance on how to formulate and conduct studies in support of coastal flooding, shore protection, and navigation projects. This Change 2 to EM 1110-2-1100, 1 April 2008, includes the following changes and updates:

    a. Part I-1. References were checked and some were deleted (Engineer Manuals that are no longer in the USACE inventory).

    b. Part I-4. Minor changes were made in the text to better reflect the contents of subsequent parts of the CEM.

    c. Part II-1. Figure II-1-9 has been revised; Equations II-1-128, II-1-160, and II-1-161 have been corrected.

    d. Part II-2. Equations II-2-4, II-2-5, and II-2-32 have been corrected along with other errors reported by various users.

    e. Part II-5. References were checked and some were deleted (Engineer Manuals that are no longer in the USACE inventory).

    f. Part II-6. The value of "e" used in Eq. II-6-28 has been corrected.

    g. Part II-7. The table of contents was corrected. A new section, II-7-11, Note to Users, Vessel Buoyancy, was added at the end of the chapter.

    h. Part III-3. Corrections have been made to format and spelling. Different plots were added to Figures III-3-24 and III-3-26.

    i. Part IV-1. Corrections have been made to references.

    j. Part V-1. Citation of an Engineer Regulation has been corrected.

    k. Part V-2. Citation of references has been changed, web pages with sources of wind and wave data have been added. Some minor text changes have also been made.

    l. Part V-3. Citations of unpublished reports or personal communications have been deleted, and links to other figures or parts of the CEM have been checked and corrected.

    m. Part V-4. Minor text changes, corrections to references and Figure V-4-1.

    n. Part V-5. Links to other parts of the CEM that were planned but never written have been deleted.

2. Applicability. This manual applies to all HQUSACE elements and all USACE commands having Civil Works and military design responsibilities.

3. Discussion. The CEM is divided into five parts in two major subdivisions: science-based and engineering-based. The first four parts of the CEM and Appendix A compose the science-based subdivision:

Part I, "Introduction"
Part II, "Coastal Hydrodynamics"
Part III, "Coastal Sediment Processes"
Part IV, "Coastal Geology"
Appendix A, "Glossary"

The engineering-based subdivision is oriented toward a project-type approach, Part V, "Coastal Project Planning and Design."

4. Distribution Statement. Approved for public release, distribution unlimited.

5. Note to Users. Revised chapters are dated 1 April 2008. Readers need to download the entire new chapters and discard earlier versions in their possession.

FOR THE COMMANDER:

STEPHEN L. HILL
Colonel, Corps of Engineers
Chief of Staff

Table of Contents

Chapter I-1
Introduction

I-1-1. Purpose and Scope

The *Coastal Engineering Manual* (CEM) assembles in a single source the current state-of-the-art in coastal engineering to provide appropriate guidance for application of techniques and methods to the solution of most coastal engineering problems. The CEM provides a standard for the formulation, design, and expected performance of a broad variety of coastal projects. These projects are undertaken to provide or improve navigation at commercial harbors, harbor works for commercial fish handling and service facilities, and recreational boating facilities. As an adjunct to navigation improvements, shore protection projects are often required to mitigate the impacts of navigation projects. Beach erosion control and hurricane or coastal storm protection projects provide wave damage reduction and flood protection to valuable coastal commercial, urban, and tourist communities. Environmental restoration projects provide a rational layout and proven approach to restoring the coastal and tidal environs where such action may be justified, or required as mitigation to a coastal project's impacts, or as mitigation for the impact of some previous coastal activity, incident, or neglect. As the much expanded replacement document for the *Shore Protection Manual* (1984) and several other U.S. Army Corps of Engineers (USACE) manuals, the CEM provides a much broader field of guidance and is designed for frequent updates.

I-1-2. Applicability

This manual is applicable to U. S. Army Corps of Engineers (USACE) Commands having civil works responsibility. It is anticipated that the comprehensive scope and instructions of this manual will warrant its use by a broad spectra of coastal engineers and scientists beyond the bounds of the USACE. Although this broad application has been considered throughout the development of the CEM, some sections are specific to the mission, authority, and operation of the USACE.

I-1-3. Definitions

Definitions are listed throughout the manual when terms are first introduced. In addition, a glossary of terms is provided in the appendix, and Table IV-1-1 lists definitions of common coastal geologic features. However, a few basic definitions will help the novice to better understand and grasp the purpose and scope of the CEM. Part IV, Chapter 2 defines types of coastal structures.

   a. Coastal. Referring to the zone where the land meets the sea, a region of indefinite width that extends inland from the sea to the first major change in topography. In this manual, "coastal" will refer to shores that are influenced by wave processes (oscillatory flow dynamics). Bays, and lakes, and estuaries are included, but rivers, primarily influenced by generally unidirectional currents, are generally beyond the scope of this manual. Estuaries, including that part of rivers subject to the ebb and flow of the tide are covered by this manual.

   b. Coastal engineering. One of several specialized engineering disciplines that fall under the umbrella of civil engineering. It is a composite of many physical science and engineering disciplines having application in the coastal area. It requires the rational interweaving of knowledge from a number of technical disciplines to develop solutions for problems associated with natural and human induced changes in the coastal zone, the structural and non-structural mitigation of these changes, and the positive and negative impacts of possible solutions to problem areas on the coast. Coastal Engineers may utilize contributions from the fields of geology, meteorology, environmental sciences, hydrology, physics, mathematics, statistics, oceanography, marine science, hydraulics, structural dynamics, naval architecture, and others in developing an understanding of the problem and a possible solution. The Coastal Engineer must consider the processes present in the area of interest such as:

- Environmental processes (chemical, ecological).

- Hydrodynamics processes (winds, waves, water level fluctuations, and currents).

- Seasonal meteorological trends (hurricane season, winter storms).

- Sediment processes (sources, transport paths, sinks, and characteristics).

- Geological processes (soil and strata characteristics, stable and migrating sub-aerial and sub-aqueous features, rebounding or subsiding surfaces).

- Long-term environmental trends (sea level rise, climate change).

- Social and political conditions (land use, development trends, regulatory laws, social trends, public safety, economics).

Harbor works, navigation channel improvements, shore protection, flood damage reduction, and environmental preservation and restoration are the primary areas of endeavor.

c. Coastal science. This field is a suite of interdisciplinary technologies applied to understanding processes, environments, and characteristics of the coastal zone. Coastal Engineers use these understandings to develop physical adaptations to solve problems and enhance the human interface with the coast.

I-1-4. Bibliography

Technical and scientific literature cited in each chapter is listed in the chapter references.

I-1-5. References

The following are official USACE engineer regulations (ER), engineer manuals (EM), engineer pamphlets (EP), and technical manuals (TM) found in the bibliographies following each chapter. They are highlighted here for easy USACE use.

TM 5-850-1
Engineering and Design of Military Ports

ER 1105-2-100
Planning Guidance Notebook

EP 1165-2-1
Digest of Water Resources Policies and Authorities

EM 1110-1-1802
Geophysical Exploration for Engineering and Environmental Investigations

EM 1110-1-1804
Geotechnical Investigations

EM 1110-2-1204
Environmental Engineering for Coastal Shore Protection

EM 1110-2-1415
Hydrologic Frequency Analysis

EM 1110-2-1613
Hydraulic Design of Deep-Draft Navigation Projects

EM 1110-2-1615
Hydraulic Design of Small Boat Harbors

EM 1110-2-1906
Laboratory Soils Testing

EM 1110-2-2301
Test Quarries and Test Fills

EM 1110-2-2302
Construction with Large Stone

EM 1110-2-5025
Dredging & Dredged Material Disposal

EM 1110-2-5026
Beneficial Uses of Dredged Material

EM 1110-8-1 (FR)
Winter Navigation on Inland Waterways

I-1-6. Acknowledgments

Authors of Chapter I-1:

Joan Pope, U.S. Army Engineer Research and Development Center, Vicksburg, Mississippi.
John H. Lockhart, Jr., Headquarters, U.S. Army Corps of Engineers, Washington, DC, (retired).

Reviewer:

Andrew Morang, Ph.D., CHL

Table of Contents

List of Figures

List of Tables

Chapter I-2
Coastal Diversity

I-2-1. Introduction

The coasts, or shores, of the world are the margins separating the 29 per cent of the earth that is land from the 71 percent that is water. By reworking and often eroding the margins of the land, the seas aid streams, subsurface water, glaciers, and the wind in wearing down the continents. Sediments derived from the land are often transient along the coasts, temporarily forming beaches, bars or islands before coming to rest on the sea floor. There is significant natural diversity in shore types throughout the United States and even greater diversity throughout the world (see Part IV for details). Consequently, engineering, development, and policy strategies need to be tailored for each unique region and need to be flexible to changes in the local condition. Coastal engineers, managers, and planners need to be aware of coastal diversity for a number of reasons:

   a. The coast is dynamic and constantly evolving to a new condition.

   b. The balance and interaction of processes are different in different areas - understanding diversity provides clues to the critical factors that may affect a particular study site.

   c. Different settings imply different erosion and accretion sediment patterns.

   d. Analytical tools and procedures may be suitable for a particular setting but totally inappropriate for another.

   e. Similarly, engineering solutions may only be appropriate for certain settings where they will function properly.

Shorelines are subject to a broad range of processes, geology, morphology, and land usages. Although winds, waves, water levels, tides, and currents affect all coasts, they vary in intensity and relative significance from one location to another. Variations in sediment supply and geological setting add to this coastal diversity. A more detailed discussion and analysis of the processes at work along the United States coasts is given by Francis P. Shepard and Harold R. Wanless in their book *Our Changing Coastline* (1971).

I-2-2. Coastal Areas

The popular image of a long, straight, sandy beach with a sandy backshore and foreshore, vegetated sand dunes, and gently sloping near shore zone with rhythmic plunging breakers may be the ideal image of the zone where the land meets the sea, but is not the norm along most coasts. Not all coastal areas are sandy, nor are all shores dominated by wave action. Some coastal areas have scenic clay bluffs or rocky headlands. Others are shallow mud flats or lush wetlands. For some shores, tidal currents or river discharge dominate sediment transport and the shore character. For other shores, the effects of glaciers, marine life (coral), or volcanoes may control the geomorphology. Shore materials include transportable muds, silts, sands, shells, gravels, and cobbles, and insitu rock formations or bedrock (erosive and non-erosive). In portions of the United States, the coastal area is sinking and gradually becoming permanently inundated; in other areas, new lands are accreting or even rising out of the sea.

   a. *Atlantic North: Glaciated coast* (Figures I-2-1, I-2-2). These coasts are normally deeply indented and bordered by numerous rocky islands. The embayments usually have straight sides and deep water as a

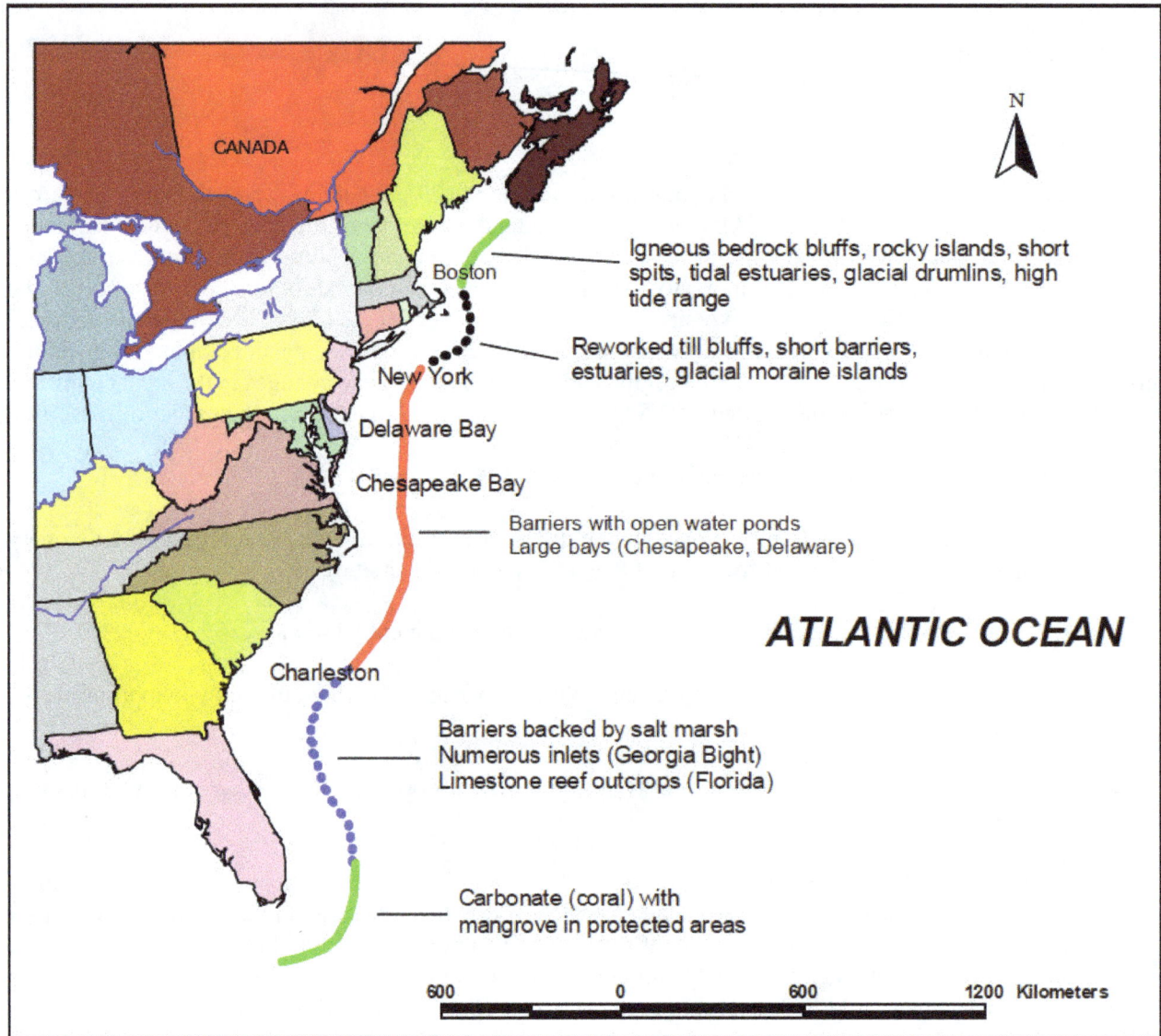

Figure I-2-1. Atlantic coast characteristics

result of erosion by the glaciers. Uplifted terraces may be common along these coasts that were formerly weighted down by ice. Abrupt changes in coastal character occur where glacial deposits and particularly glacial outwash play a dominant role, while in some rocky areas, few glacial erosion forms can be found. Moraines, drumlins, and sand dunes, the result of reworking outwash deposits, are common features. Glaciated coasts in North America extend from the New York City area north to the Canadian Arctic (Figures I-2-3, I-2-4, IV-2-8, and IV-2-9), on the west coast, from Seattle, Washington, north to the Aleutian Islands, and in the Great Lakes. (Figure IV-2-20) (Shepard 1982).

b. Atlantic Central and South: Barrier and drowned valley coasts. South of the glacial areas begins the coastal Atlantic plain, featuring almost continuous barriers interrupted by inlets and by large embayments with dendritic drowned river valleys, the largest being Delaware and Chesapeake Bays. The North American coastline is reported to include over 10,000 km of barriers, about 33 percent of all barrier coast of the world (Berryhill, Dixon, and Holmes 1969). The United States alone has a total length of 4,900 km of barriers and spits, the longest extent for a single nation (Figure I-2-5 and Table IV-2-3). Extensive wetlands and marshes

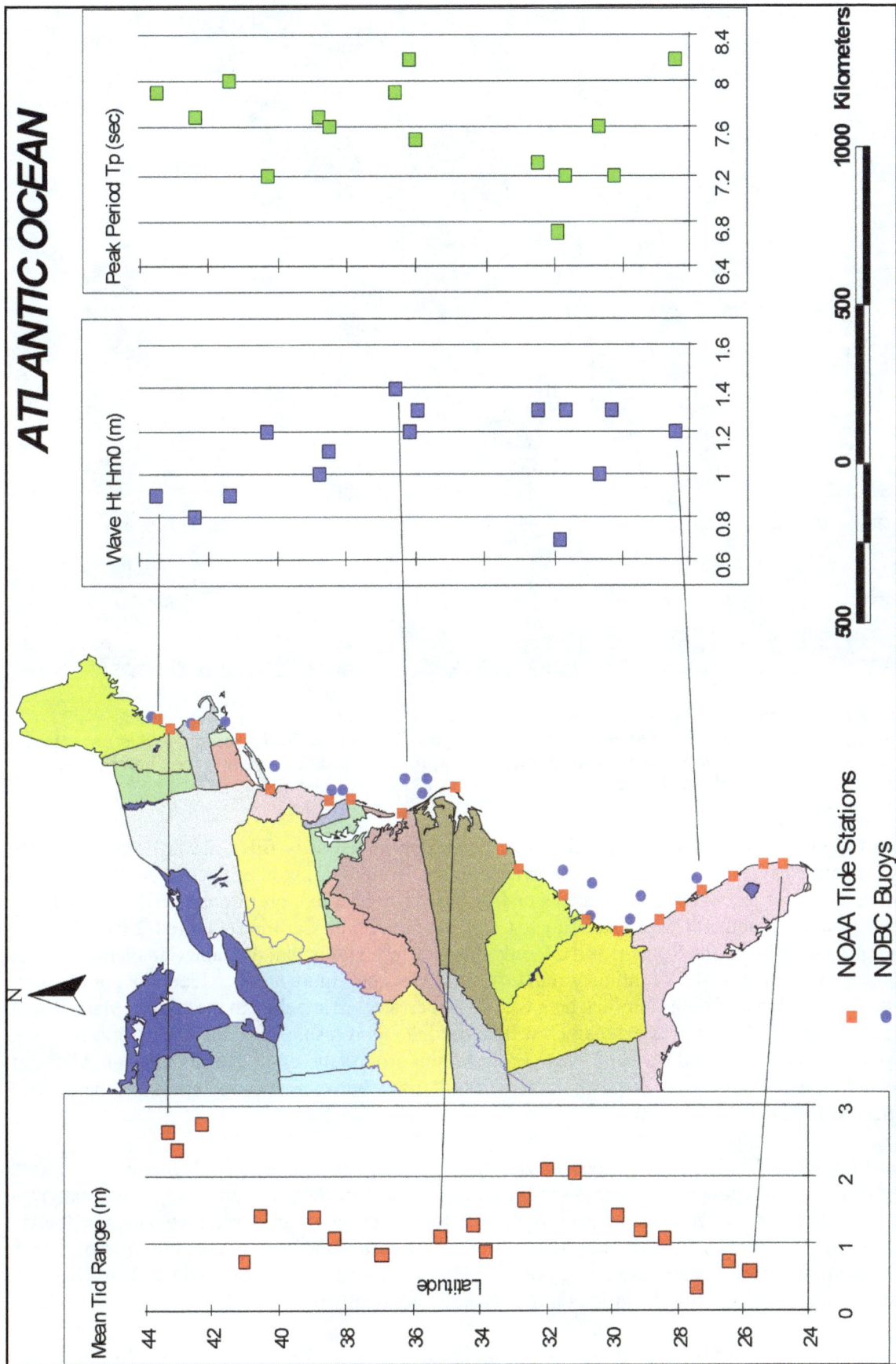

Figure I-2-2. Tide and wave characteristics of the Atlantic coasts. Wave data summarized from National Data Buoy Center buoys. H_mO and T_p averaged from hourly statistics over total period of record from statistics computed by National data Buoy Center. Tide range for indicated stations from statistics presented in NOAA Tide Tables

Figure I-2-3. Barrier Island and bay complex, southern Rhode Island. View looking west toward Quonochontaug Point, a rocky headland with bedrock outcrops. The barrier in the foreground is East Beach, with Block Island Sound to the left and Ninigret Pond to the right. Prominent overwash fans can be seen in the shallow waters of the pond (April 1977)

mark much of the coast, where sediment and marsh vegetation have partly filled the lagoons behind the barriers. Some coasts have inland ridges of old barrier islands, formed during interglacial epochs, separated from the modern barrier islands by low marshes or lagoons. The best exhibit of cuspate forelands in the world extends from the mouth of Chesapeake Bay to Cape Romain, South Carolina (Figure I-2-6). The coast is much straighter south of Cape Romain and the only cuspate foreland is that of Cape Canaveral, Florida. Barrier Islands and drowned valleys continue south to Miami, Florida (Figure I-2-7), except for a brief length of coast in the Myrtle Beach, South Carolina, area where the barriers are attached to the coastal plain. Much of the southeast coast of Florida was extensively filled, dredged, and reshaped in the early 20th century to support development (Lenček and Bosher 1998). From Miami around the tip of Florida through Alabama, Mississippi and eastern Louisiana, coastal characteristics alternate between swampy coast and white sand barriers (Shepard 1982).

c. The Atlantic and Gulf of Mexico: Coral and mangrove coasts. The barrier islands change from quartz sand south of Miami to carbonate-dominated sand, eventually transforming into coral keys and mangrove forest. The Florida Keys are remnants of coral reefs developed during a higher sea level stage of the last interglacial period. Live reefs now grow along the east and south side of the keys and the shallows of Florida Bay studded with mangrove islands extending north and west into the Everglades and the Ten Thousand Islands area that comprises the lower Florida Gulf of Mexico coast (Shepard 1982).

Figure I-2-4. New York Harbor, late 1930s. This drowned river valley system, partly sculpted by glaciers, is one of the world's finest natural harbors. The USACE has an active role dredging, clearing debris, and maintaining navigability of this great port. View looking north, with Manhattan in the center and Brooklyn to the right. Photograph from Beach Erosion Board archives

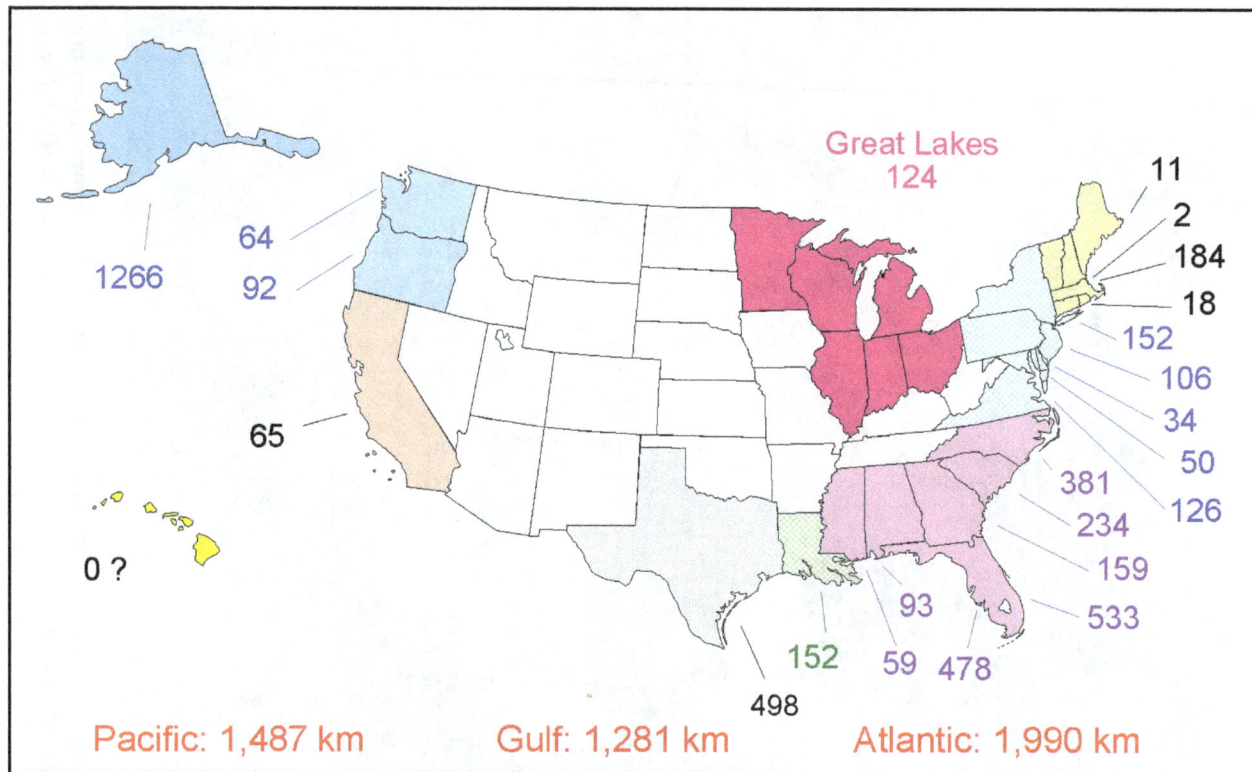

Figure I-2-5. Length (in km) of barrier islands and spits in the United States. Data measured from U.S. Geological Survey topographic maps (see Table IV-2-3 for details)

d. Gulf of Mexico East: Wetland mangrove, and barrier coasts (Figures I-2-8, I-2-9). On Florida's Gulf of Mexico coast, barrier islands begin at Cape Romano and extend north as far as Cedar Keys. Enclosed bays usually have an abundance of mangrove islands and the topography is low with many lakes and marshes. North of Cedar Keys, the barrier islands end. They are replaced by a vast marsh doted with small vegetated islands. The rock strata in this area are limestone, which, along with the low river gradients and numerous ponds or sinkholes, accounts for the absence of sand in the region. Due to its location and the large shallow water area offshore, little wave energy is present except during rare hurricanes. Some 130 km to the northwest, the swamp coast ends. Here the coastal trend changes direction from north-south to east-west, and Ochlockonee Bay, with drainage from the southern Appalachian Mountains, provides quartz sand for redevelopment of barrier islands. These sandy islands, with their various openings for access to the lowland port cities, continue westward as far as the Mississippi River delta (Figures I-2-10 and I-2-11).

Studies of the Mississippi delta indicate that the river has built a series of deltas into the Gulf of Mexico during postglacial times and that the Balize Delta (bird foot) is the latest, with an age of about 1500 years. The Bird Foot delta is southeast of New Orleans, lying among a series of old passes that extend for 300 km (186 miles) along the coast. Most of the greater Mississippi delta is marshland and mud flats, with numerous shallow lakes and intertwining channels (Figures I-2-12 and IV-3-9). The principal rivers have built natural levees along their course. These natural levees are about a meter above the normal water level, but many of them have been artificially raised to provide protection to towns and cities from floods. Aquatic plants cover the marshland, which is remarkable for the huge population of waterfowl it supports. In the areas of old delta lobes, subsidence has left only the natural levees above water in some instances.

e. Gulf of Mexico West: Barrier coast. From western Louisiana, west of the Mississippi Delta marsh coast, toward the southwest, barrier islands become the dominant coastal features. Some of the longest

Figure I-2-6. Cape Hatteras, North Carolina, view north. The Atlantic Ocean is to the right, and the bay to the left of the barrier is Pamlico Sound. The rough water in the foreground is the infamous Diamond Shoals, known as the "Graveyard of the Atlantic." The bump in the shoreline is the location of the Cape Hatteras lighthouse, which was recently moved inland away from the receding shore. A mature maritime forest has grown on the beach ridges in the central portion of the barrier. The forest indicates that this portion of the island has been stable for several hundred years. Photograph taken February 28, 1993, during the waning stage of an extratropical storm

barrier islands in the world are located along the Texas coast. Padre Island and Mustang Island, combined, extend for 208 km and feature extensive dune fields behind the broad beaches. The dunes rarely rise more than 10 m in height, and many marshy wash-over deltas have extended into the large lagoons behind the barriers. The lagoons and estuaries decrease in depth toward Mexico. A large part of Laguna Madre is only inundated during flood periods or when the wind blows water from Corpus Christi Bay onto the flats. River deltas are responsible for much of this infilling, resulting in large differences between recent chart depths and those of 100 years ago (Shepard 1982).

f. Pacific: Sea cliffs and terraced coasts (Figures I-2-13, I-2-14). Low sea cliffs bordered by terraces and a few coastal plains and deltas compose the coasts of southern California. Blocks form projections into the sea and feature a series of raised terraces such as those at Point Loma, Soledad Mountain, and the San Pedro Hills in the Los Angeles area. North of Los Angeles, the Santa Monica Mountains follow the coast. Sea cliffs in this area are actively eroding, particularly in areas where they have been cut into alluvium (Figure I-2-15). At Point Conception, the coast trends north-northwest and a different

Figure I-2-7. Hallandale Beach, an example of a popular recreation beach in an urban area on the Atlantic coast of southeast Florida. Photograph taken June 27, 1991, after the beach had been renourished using sand hydraulically pumped from an offshore source. Stakes in the beach were used as survey markers

geomorphology is evident. Despite the presence of a series of regional mountain ranges that cut across the coast, the rugged central and northern California coast is one of the straightest in the world. This area has high cliffs with raised marine terraces. A few broad river valleys interrupt the mountainous coast. Here, river sediments have been returned by the waves to the beaches and carried inland by westerly winds to form some unusually large dune fields. Monterey and San Francisco Bays, the two largest embayments, are at the mouths of the Salinas and the San Joaquin-Sacramento rivers respectively; the latter drains the great central valley of California. North of Cape Mendocino, the coast trends almost directly north, through Oregon and Washington, to the Strait of Juan de Fuca. Along this coast, lowland valleys at the mouths of large rivers alternate with short, relatively low mountainous tracts. Barriers or spits have formed at river mouths, as have large dune fields (Figure I-2-16). Many of the rivers, including the great Columbia, discharge into estuaries. This indicates that the rivers have not yet been able to fill drowned valleys created by the sea level rise when the great Pleistocene continental glaciers melted (Shepard 1982).

Because of the North Pacific Ocean's harsh wave climate, all of the major cities in Oregon and Washington were founded in sheltered water bodies. For example, Vancouver, Washington, and Portland, Oregon, are on the Columbia River. Puget Sound, a deep, sheltered, fjord-like water body in western Washington State, provides safe access for ships steaming to Tacoma, Bellingham, Everett, and Seattle (Figure I-2-17).

g.   The Bering and Chukchi Seas: Arctic coastal plains and barriers (Figure I-2-18). The volcanic Aleutian Mountains trend southwest from Anchorage, Alaska, to form the Alaska Peninsula and the Aleutian Islands that extend some 2200 km (1370 miles) forming the border between the Pacific Ocean and the Bering Sea (Figure I-2-14). Beyond the Alaska Peninsula and bordering the Bering Sea, extensive coastal plains

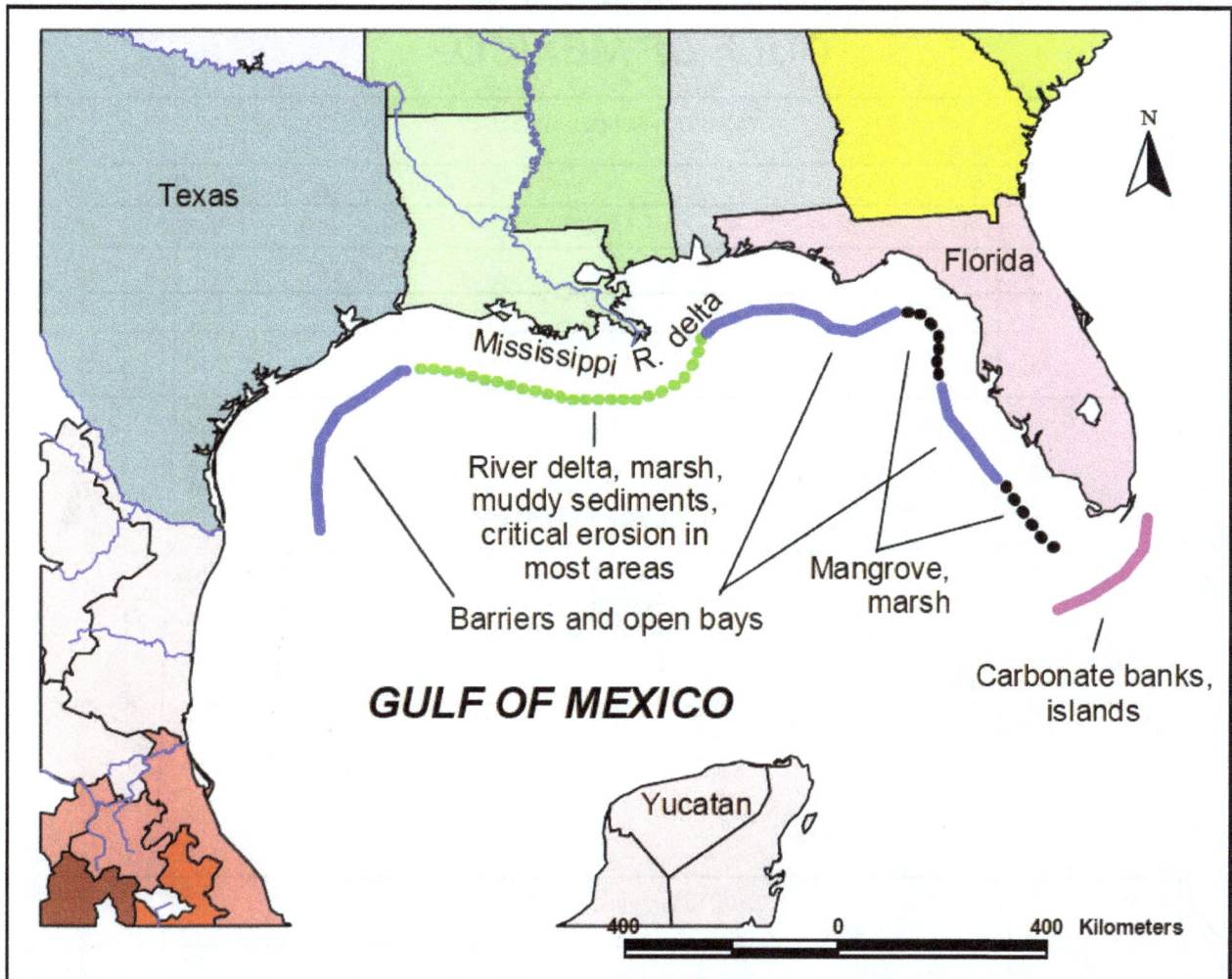

Figure I-2-8. Gulf of Mexico coastal characteristics

are found with numerous lakes and meandering streams. Only a few mountain ranges extend as points into the sea. The Yukon River has formed a large delta with many old lobes that form a vast plain connecting small, elevated tracts. The oldest is located in the now drowned mouth of the Kuskokwim River. One reason this coast differs from the glaciated southern coast of Alaska, is because it was largely ice- free during the Pleistocene era. Permafrost becomes more important to the north where it greatly increases the number of surface depressions in the summer when it melts forming *thaw lakes*. Rising above the coastal plain with mountains over 1,000 m, the Seward Peninsula with Norton Sound and the Bering Sea to the south and Kotzebbue Sound and the Chukchi Sea to the north provides a great contrast to the adjoining coasts. North of Kotzebue Sound, barriers and cuspate forelands similar to those of North Carolina border the coast. The first cuspate foreland is the unusual Point Hope. Three more cuspate forelands extend along the coast terminating with Point Barrow, the most northern point of Alaska (Shepard 1982).

h. The Beaufort Sea: Deltaic coast. East of Point Barrow, the coast is dominated by river deltas. Rivers draining the Brooks Range and father east the Mackenzie, draining the northern Canadian Rockies, built these deltas even though the rivers flow only a short period each year. Where the deltas are not actively building into the sea, extensive barrier islands can be found (Shepard 1982). One of the dominant processes in shaping beaches in Alaska is the ride-up of shore ice (Kovacs 1983).

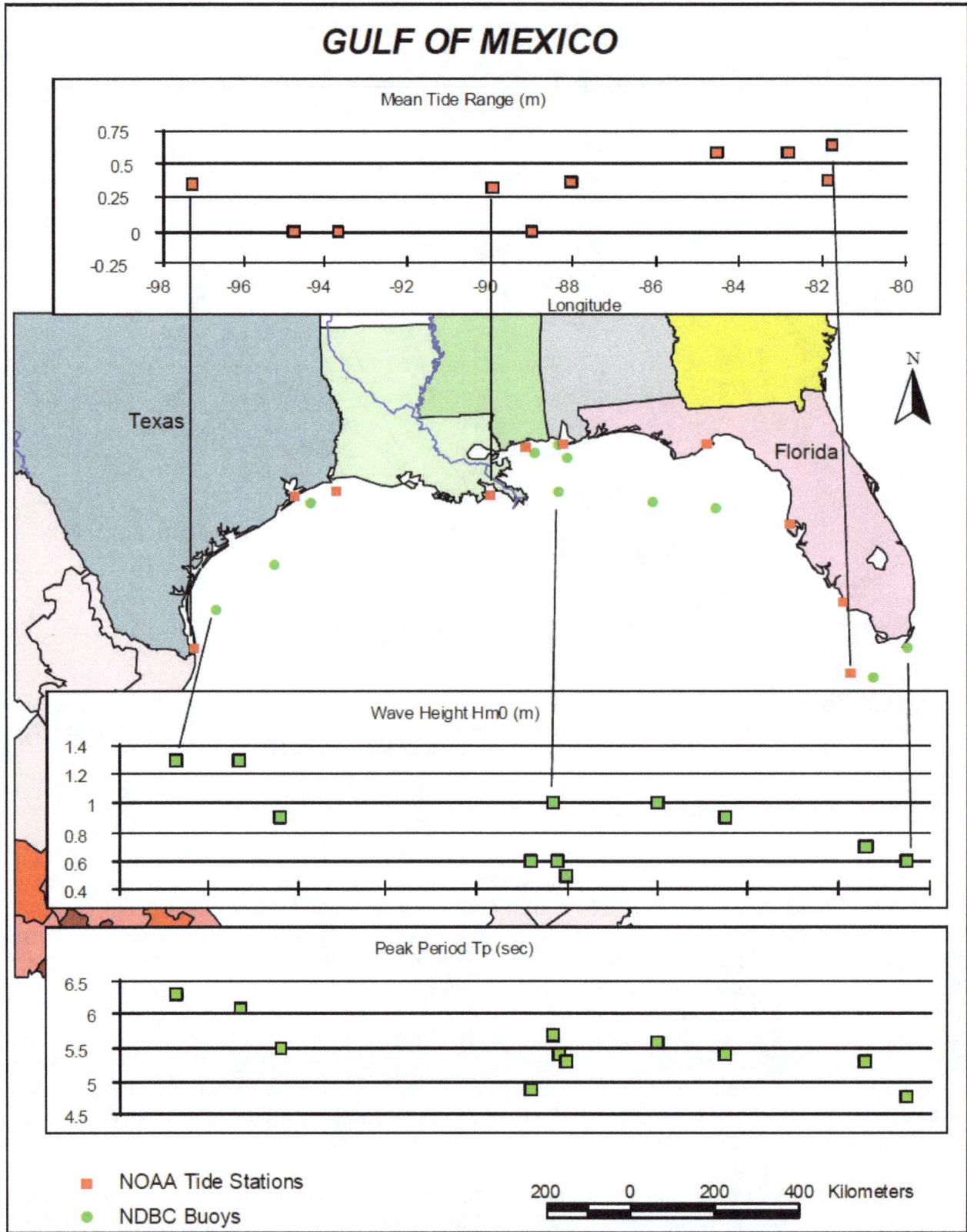

Figure I-2-9. Tide and wave characteristics of the Gulf Coast

Figure I-2-10. East Pass Inlet, Florida, View looking west towards Santa Rosa Island, with the Gulf of Mexico on the left and Choctawhatchee Bay to the right. The barrier island immediately beyond the inlet is part of Eglin Air Force Base and has remained undeveloped. The beach in the foreground is Holiday Isle, which has been heavily commercialized. This area of Florida is noted for its brilliant white quartz sand and excellent fishing. The inlet is a Federal navigation project with converging rubble-mound jetties. Photograph taken March 1991

i. Pacific: Volcanic islands (Figure I-2-19). The Hawaiian archipelago extends from the large island of Hawaii across the central Pacific Ocean northwest to tiny Kure Atoll, 2450 km away. The eight main islands of the state of Hawaii, at the southeast end of the archipelago comprise 99 percent of the land area. About 20 percent of the 1,650 km of shore on the main islands is sandy beach (USACE 1971). Aside from manmade structures, the remainder of the shore consists primarily of outcrops or boulders of lava, but also includes muddy shores, gravel beaches, beach rock, raised reefs, and lithified sand dunes. Elevations of the rocky shores vary from 1-2 m high raised reefs to 600 m sea cliffs along the Napali coast of Kauai. The Hawaiian Islands are the tops of volcanic mountains rising above the ocean floor about five km below the water surface. These volcanoes formed over a localized hot spot of magma generation. As the older volcanoes formed great shields and died, the movement of the ocean floor and crust moved them to the northwest. A higher percentage of sand shores are found on the older islands, see Table I-2-1. Beaches on Hawaiian Islands are smaller than those on the continental shores, because of the young age of the islands, the absence of large rivers to supply sediment, and the shape and exposure of the island beaches to the wave systems that affect the islands. The sand on the beaches is also different in that it is primarily calcareous and of biologic origin. The calcareous sand originates as shells and test of animals or algae that live on the fringing reefs or shallow waters adjacent to the islands. Two exceptions are some beaches near stream mouths are detritus basalt sand, and a few beaches on the island of Hawaii are black volcanic glass sand generated by the steam explosions that occur when hot lava flows into the ocean (Moberly and Chamberlain

Figure I-2-11.   Morgan Peninsula, Alabama, about 10 km east of the mouth of Mobile Bay (21 April 1998). This is the back side of the barrier island, with Mobile Bay in the right side of the photograph. The dead trees clearly show that the shore has retreated within the last few years. In this portion of the Alabama shore, erosion on the back side of the barrier is a more serious threat than on the ocean side

Table I-2-1
Age and Sandy Shores of the Major Hawaiian Islands

Island	Age (million years)	Total Shoreline (km)	Sandy Shoreline (km)	Percent Sandy Shoreline
Kauai	5.1	182	80	44
Oahu	3.7 to 2.6	319	90	28
Molokai	1.5 to 1.9	170	40	24
Lanai	NA	84	29	35
Maui	1.3 to 0.9	256	54	21
Hawaii	Active	492	35	7

Based on Campbell & Moberly 1985.

Figure I-2-12. Dulac, Louisiana (March 1981). Located near the Gulf of Mexico entrance to the Houma Navigation Channel, many residents of Dulac and other towns in the Acadian parishes of southern Louisiana depend on the water for their livelihoods - shrimping, fishing, and servicing the offshore petroleum industry. Although about 25 km from the Gulf, Dulac, at an elevation of 1-2 m above sea level, is highly vulnerable to hurricanes and flooding

1964). The coastal geology of each island is derived from the erosion of the island shield and subsequent volcanic activity (Campbell and Moberly 1985).

   j. Great Lakes of North America. The five Great Lakes, Ontario, Erie, Huron, Michigan, and Superior, are located along the Canadian and U.S. boundary, except that Lake Michigan is totally within the United States (Figure I-2-20). They have a combined surface area of 245,300 km² (94,700 miles²), making them the largest freshwater body in the world. Together with the Saint Lawrence Seaway, they form a major shipping artery that is navigable inland for 3,770 km from the Atlantic by ocean-going vessels, except from about December through April when shipping is blocked by ice (Figures I-2-21, I-2-22, and I-2-23). The lakes range in elevation from about 183 m for Lake Superior (International Great Lakes Datum 1985) to about 75 m for Lake Ontario, with the largest drop in elevation, 51 m between Lakes Erie and Ontario at Niagara Falls (CCEE 1994). Geologically, the Great Lakes are relatively young, having been formed by glacial action during the Pleistocene period. Prior to the glacial age, the area occupied by Lake Superior was a broad valley and the area occupied by the other lakes was a spreading plain. During the ice period, glaciers deepened the bed of Lake Superior and gouged deep depressions forming the beds of the other lakes. As the ice sheet retreated, fingers of ice remained in the depressions, rimmed by glacial moraines and outwash plains. Lakes were formed when the ice melted. Successive advances and retreats of the ice caps changed the drainage of the lake region until about 10,000 years ago. Then, the northern part of the area up warped or rebounded causing the lakes to drain into the St. Lawrence through what is now the Niagara River.

The shores of the Great Lakes and other freshwater lakes in the United States and throughout the world are as diverse as the ocean shores, featuring high and low erosive and non-erosive cliffs and bluffs, low plains, sandy beaches, dunes, barriers and wetlands (Figure I-2-24).

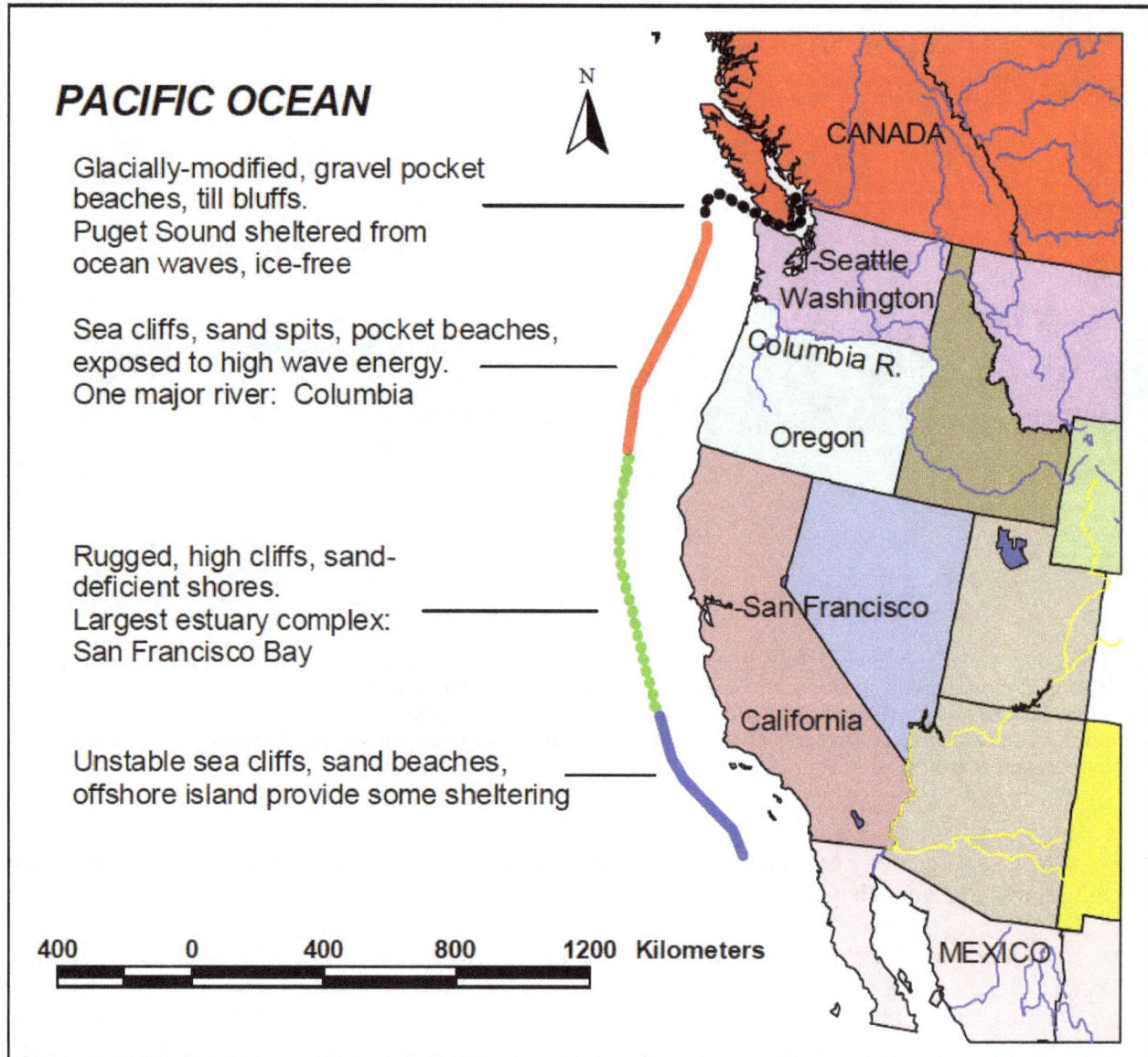

PACIFIC OCEAN

Glacially-modified, gravel pocket beaches, till bluffs.
Puget Sound sheltered from ocean waves, ice-free

Sea cliffs, sand spits, pocket beaches, exposed to high wave energy.
One major river: Columbia

Rugged, high cliffs, sand-deficient shores.
Largest estuary complex: San Francisco Bay

Unstable sea cliffs, sand beaches, offshore island provide some sheltering

CANADA
Seattle
Washington
Columbia R.
Oregon
San Francisco
California
MEXICO

400   0   400   800   1200 Kilometers

Figure I-2-13.   Pacific coastal characteristics

I-2-3. Stability

Not all shores are in equilibrium with the present littoral processes. Shores with a character inherited from previous non-littoral processes (i.e., glacial or river deposited materials) maybe doomed to significant rates of erosion under present conditions, such as the Mississippi delta of Louisiana and portions of the Great Lakes. Some shores exhibit short-term seasonal or episodic event-driven cyclic patterns of erosion and accretion (e.g., the southern U.S. Atlantic coast). Other shores demonstrate long-term stability due to balanced sediment supply and little relative sea level rise influence, such as the west coast of Florida. For some shores, very little beach-building material is available, and what little is available may be prone to rapid transport, either alongshore or offshore (e.g., the Great Lakes). Shores that have been heavily modified by man's activities usually require a continuing commitment to retain the status quo. Prime examples are New Jersey, which was extensively modified during the 20th century and is now undergoing several major

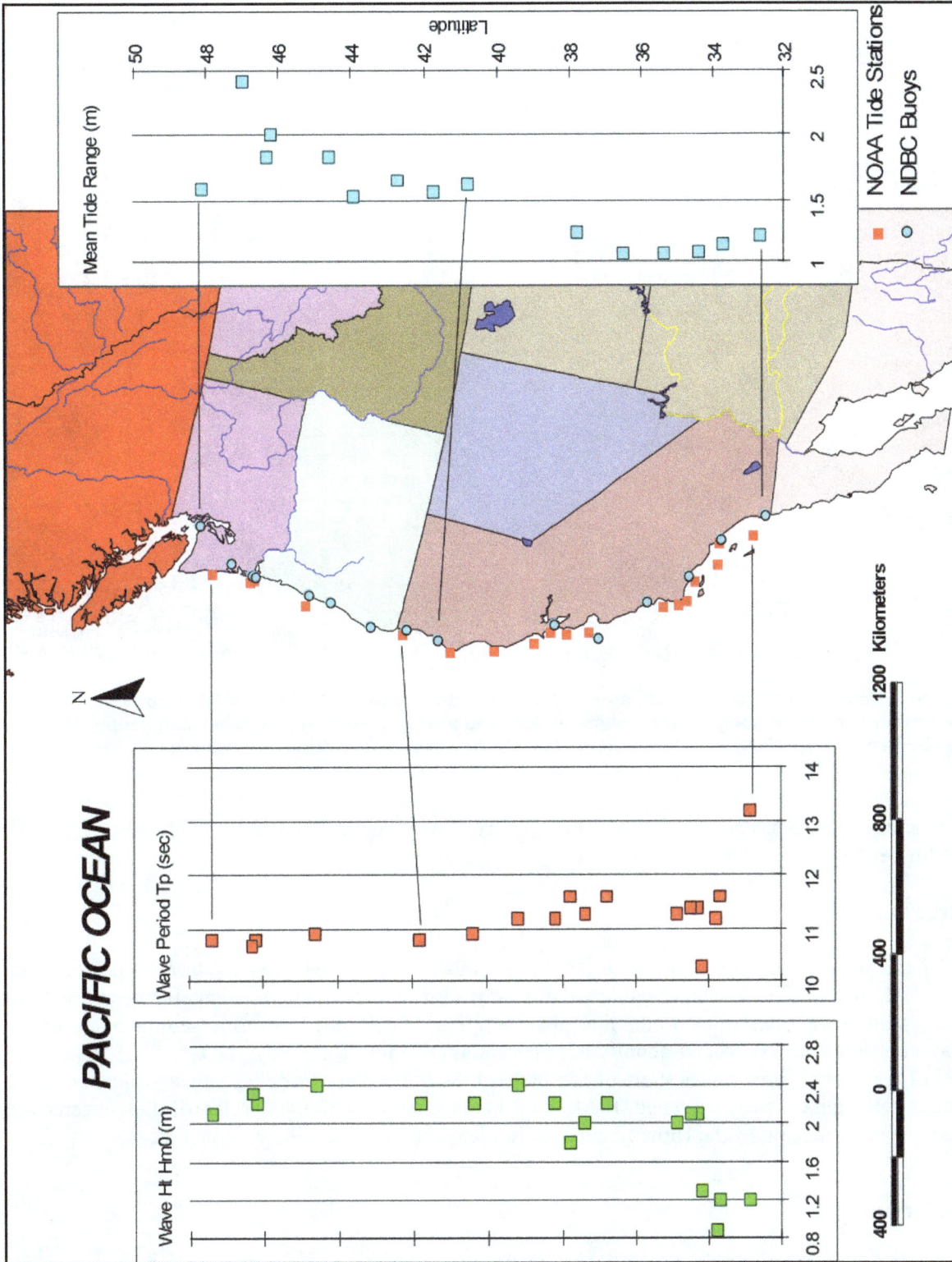

Figure I-2-14. Pacific coast tide and wave characteristics. The southernmost buoy shows high wave period because of the influence of swell waves and sheltering from wind waves provided by offshore islands

Figure I-2-15.   Pocket beach just north of Laguna Beach, southern California (April 1993).   Poorly consolidated sandstone and conglomerate bluffs in this area are highly vulnerable to erosion, jeopardizing exclusive residential properties.   Erosion is caused by storm waves and groundwater runoff

beach fills, and numerous urban areas around the country (Los Angeles, New York, Galveston, Chicago, Miami, Palm Beach).

I-2-4. Erosion

In order for one shore to accrete, often some other shore must erode. Erosion is a natural response to the water and wind processes at the shore, but erosion is only a problem when human development is at risk. Sometimes, man-made alterations to the littoral system, including modifications to sediment sources or sinks, may contribute to the eroded condition. The National Shoreline Study (DOA 1971) found that 24 percent of the entire United States shore of 135,000 km (84,000 miles) is undergoing significant erosion where human development was threatened. If Alaska, with its 24,800 km. (15,400 miles) of shore is removed from the statistic, 42 percent of the United States shore is experiencing significant erosion!

I-2-5. Solutions

There are no absolute rules, nor absolute solutions to the problem of coastal erosion given the dynamic and the diverse character of the shoreline. No single set of regulations, or single land use management philosophy, is appropriate for all coastal situations or settings. The diversity of the coasts requires consideration of a variety of solutions when addressing problems in a particular area. Solutions can be classified into five broad functional classes of engineering or management, as listed in Table I-2-2. These options are explored in detail in Part V of the CEM.

Figure I-2-16.   Mouth of the Siuslaw River, southern Oregon near the town of Florence (December 1994; view looking south).   This and other Federal navigation projects on the Oregon and Washington coasts are difficult and expensive to maintain because of high wave energy and a short construction season.   The scale of these Pacific projects is difficult to appreciate from aerial photographs: the Siuslaw rubble-mound jetties, first built in 1917, are 180 m apart and the north jetty is 2300 m long. The shore in this area consists of long barrier spits interrupted with rocky headlands

Table I-2-2
Alternatives for Coastal Hazard Mitigation

Functional Class	Approach Type
1. Armoring structures	Seawall
	Bulkhead
	Revetment - revetment
2. Beach stabilization structures and facilities	Breakwaters (including artificial headlands)
	Groins
	Sills vegetation
	Groundwater drainage
3. Beach restoration	Beach nourishment
	Sand passing
4. Adaptation and accommodation	Flood proofing
	Zoning
	Retreat
5. Combinations	Structural and restoration
	Structural and restoration and adaptation
6. Do nothing	(no intervention)
Abbreviated from CEM Part V, Table V-3-1	

Figure I-2-17. Seattle, located in sheltered Puget Sound, is one of the world's great natural anchorages. In the 1800s Seattle was a timber town and point of embarkation for Alaska and the Orient. During the 1980s and 1990s, the port has prospered with container traffic and the export of grain and other agricultural products. Areas of the harbor need regular dredging. (Photograph July 1995)

Sometimes the solutions require the use of "hard" static structures built of rock, steel, or concrete, and sometimes the solutions involve "soft" dynamic approaches, such as adding littoral material or modifying the vegetation. Chapter V-3, "Shore Protection Projects" provides a more detailed discussion of the options and limitations available to the coastal engineer.

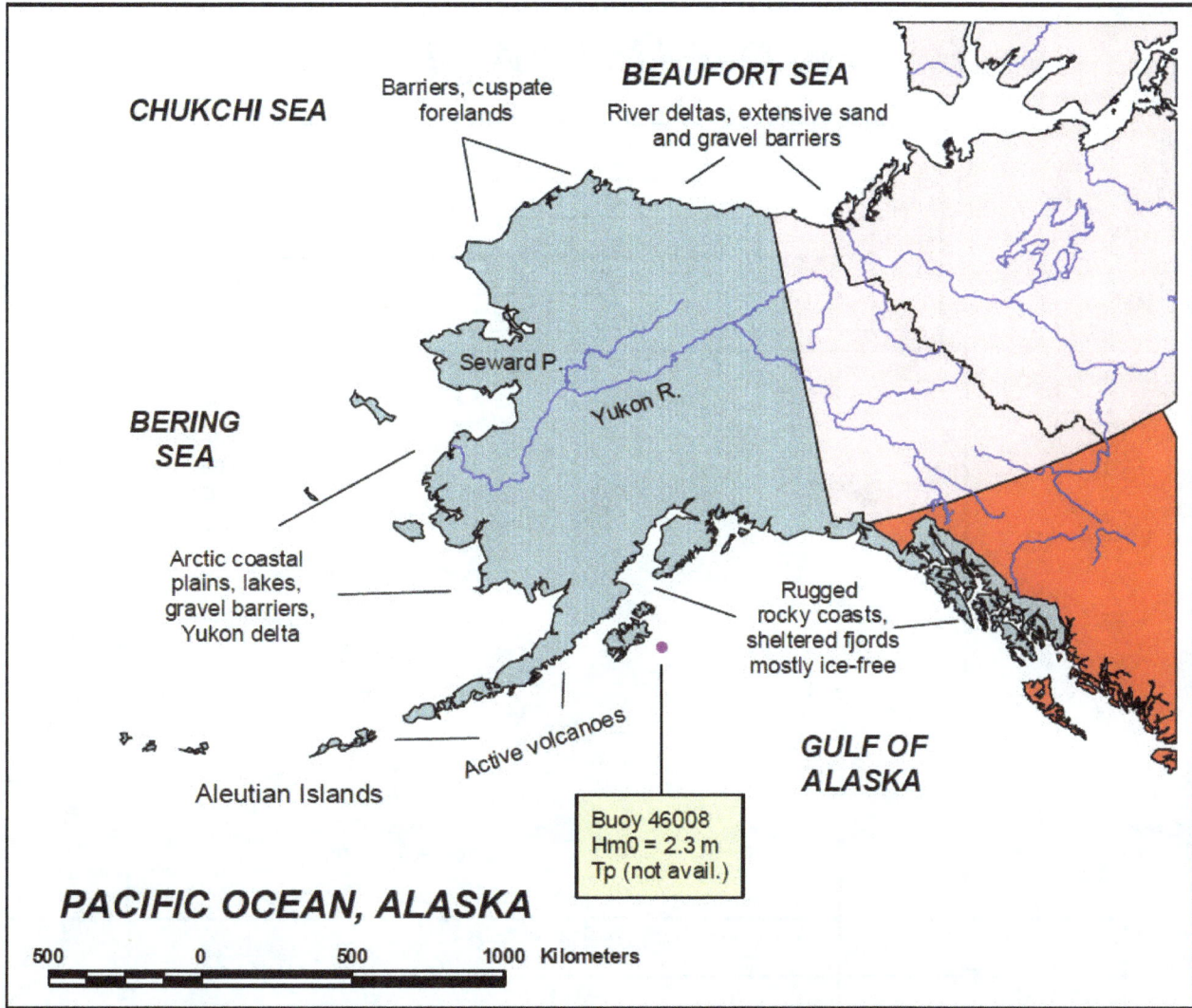

CHUKCHI SEA

Barriers, cuspate forelands

BEAUFORT SEA

River deltas, extensive sand and gravel barriers

Seward P.

Yukon R.

BERING SEA

Arctic coastal plains, lakes, gravel barriers, Yukon delta

Rugged rocky coasts, sheltered fjords mostly ice-free

Active volcanoes

Aleutian Islands

GULF OF ALASKA

Buoy 46008
Hm0 = 2.3 m
Tp (not avail.)

PACIFIC OCEAN, ALASKA

500   0   500   1000   Kilometers

Figure I-2-18. Alaska coastal characteristics

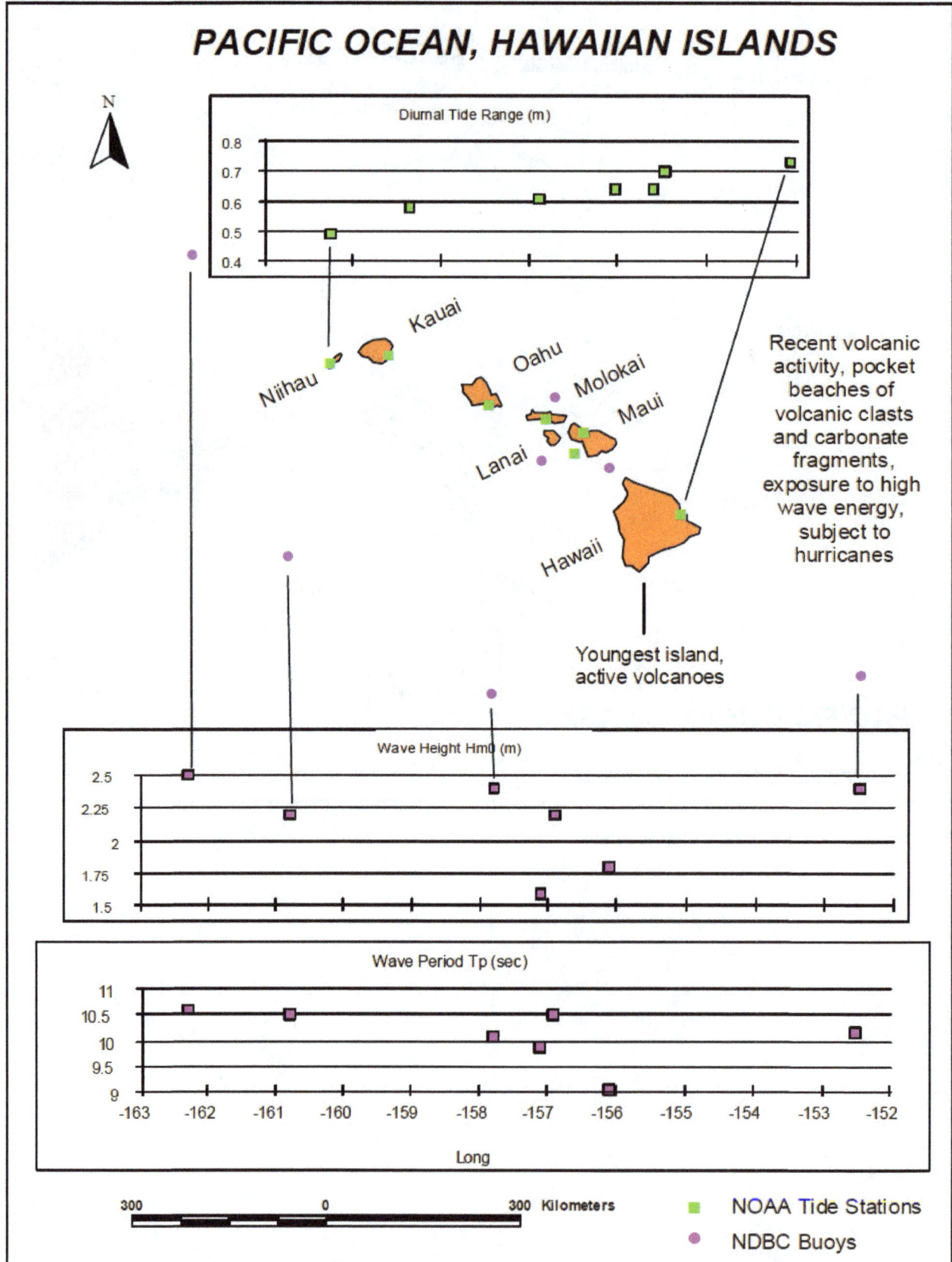

Figure I-2-19. Hawaiian Islands wave characteristics

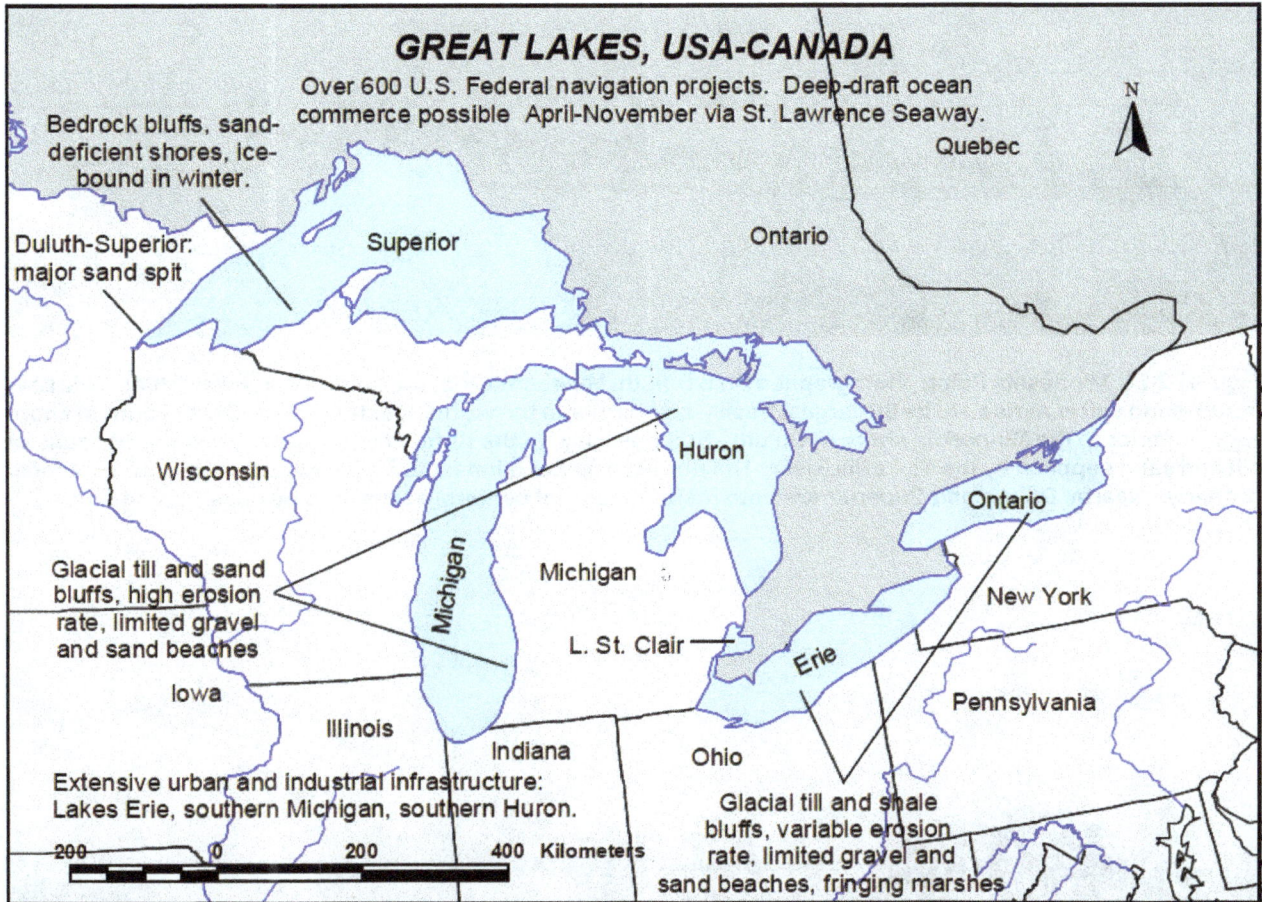

GREAT LAKES, USA-CANADA

Over 600 U.S. Federal navigation projects. Deep-draft ocean commerce possible April-November via St. Lawrence Seaway.

Bedrock bluffs, sand-deficient shores, ice-bound in winter.

Duluth-Superior: major sand spit

Glacial till and sand bluffs, high erosion rate, limited gravel and sand beaches

Extensive urban and industrial infrastructure: Lakes Erie, southern Michigan, southern Huron.

Glacial till and shale bluffs, variable erosion rate, limited gravel and sand beaches, fringing marshes

Superior

Quebec

Ontario

Wisconsin

Huron

Ontario

Michigan

New York

Michigan

L. St. Clair

Erie

Iowa

Pennsylvania

Illinois

Indiana

Ohio

200   0   200   400 Kilometers

Figure I-2-20. Great Lakes shoreline characteristics

Figure I-2-21. Minnesota Point, photographed from Duluth, Minnesota, looking south (November 1994). This bay-mouth sand spit is reputed to be the largest fresh water barrier in the world. It extends from the Wisconsin shore near Superior to the Minnesota shore at Duluth. St. Louis Bay, to the right, needs regular dredging because of silt and sand supplied by the St. Louis River. The northern part of Minnesota Point is developed with residential property. Nearby Duluth and Superior are both major industrial centers, accessible by ocean-going ships

Figure I-2-22. Calumet Harbor, Indiana (September 1985). This is an example of the industrial infrastructure found in many of the Great Lakes cities that thrived from the 1800s until the 1970s. Many of these steel mills are now closed, but some of the sites are being redeveloped for other purposes. Calumet is a Federal navigation project. The concrete cap on the breakwater in the foreground has shifted, indicating some damage to the underlying wood crib (originally built in the 1890s)

Figure I-2-23. Duluth Canal, Minnesota (November 1994). Thanks to the St. Lawrence Seaway and a network of locks, rivers, and canals, deep-draft ocean-going freighters can ship bulk commodities and goods throughout the Great Lakes. This vessel is taking iron ore from the nearby Mesabe Iron Range to some distant port. The Duluth Canal, excavated privately in 1871, is now a Federal project maintained by the USACE.

Figure I-2-24.   Bluffs about 1 km north of St. Joseph Harbor, eastern Lake Michigan (November 1993). In this area, the sand and clay bluffs are receding at an average rate of between 0.3 and 0.4 m per year. They are highly vulnerable to ground water seepage and, during periods of high lake level, to wave attack. Freshly-slumped clay blocks can be seen on the bluff face in the right side of the image

I-2-6. References

Berryhill, Dickinson, and Holmes 1969
Berryhill, H. L., Dickinson, K. A., and Holmes, C. W. 1969. "Criteria for Recognizing Ancient Barrier Coastlines," *American Association of Petroleum Geologist Bulletin 53,* pp 703-707

Campbell and Moberly 1985
Campbell, J. F., and Moberly, R. 1985. "130. Hawaii," Bird, E. C. F., and Schwartz, M. L., eds., *The World's Coastline,* Van Nostrand Reinhold, New York, NY., p 1011-1022.

CCEE 1994
Concise Columbia Electronic Encyclopedia (CCEE). 1994. Columbia University Press, Encyclopedia.com— Results for Great Lakes.

DOA 1971
Department of the Army, Corps of Engineers. 1971. *Report on the National Shoreline Study,* Washington, DC. 20314, August 1971, 59 p.

Kovacs 1983
Kovacs, A. 1983. Shore ice rid-up and pile-up features, Part I: Alaska's Beaufort sea coast, CRREL Report CR-83-9-PT-1, U.S. Army Cold Regions Research and Engineering Laboratory,Hanover, NH.

Lenček and Bosker 1998
Lenček and Bosker. 1998. *The Beach: The History of Paradise on Earth.* Viking New York, 310 p.

Moberly and Chamberlain 1964
Moberly, R. and Chamberlain, T. 1964. "Hawaiian Beach Systems," *Hawaii Institute of Geophysics Report 64-2,* University of Hawaii.

Shepard and Wanless 1971
Shepard, Francis P., and Wanless, Harold R. 1971. *Our Changing Coastlines,* McGraw-Hill, New York, NY., 579 p.

Shepard 1982
Shepard, Francis P. 1982. "North America, Coastal Morphology." *Encyclopedia of Beaches and Coastal Environments, Encyclopedia of Earth Sciences,* Volume XV, Schwartz, Maurice L. Editor, Hutchinson Ross Publishing Co., 940 p, pp 593-603.

I-2-7. Acknowledgments

Authors of Chapter I-2, "Coastal Diversity:"

John H. Lockhart., Jr., Headquarters, U.S. Army Corps of Engineers, Washington, DC, (retired).
Andrew Morang, Ph.D., Coastal and Hydraulics Laboratory (CHL), Engineer Research and Development
Center, Vicksburg, Mississippi.

Reviewer:

Joan Pope, CHL

Table of Contents

List of Figures

Chapter 1-3
History of Coastal Engineering

I-3-1. Ancient World

The history of coastal engineering reaches back to the ancient world bordering the Mediterranean Sea, the Red Sea, and the Persian Gulf. Coastal engineering, as it relates to harbors, starts with the development of maritime traffic, perhaps before 3500 B.C. Shipping was fundamental to culture and the growth of civilization, and the expansion of navigation and communication in turn drove the practice of coastal engineering. The availability of a large slave labor force during this era meant that docks, breakwaters, and other harbor works were built by hand and often in a grand scale similar to their monumental contemporaries, pyramids, temples, and palaces. Some of the harbor works are still visible today, while others have recently been explored by archaeologists. Most of the grander ancient harbor works disappeared following the fall of the Roman Empire. Earthquakes have buried some of the works, others have been submerged by subsidence, landlocked by silting, or lost through lack of maintenance. Recently, archaeologists, using modern survey techniques, excavations, and old documents, have revealed some of the sophisticated engineering in these old harbors. Technically interesting features have shown up and are now reappearing in modern port designs. Common to most ancient ports was a well-planned and effectively located seawall or breakwater for protection and a quay or mole for loading vessels, features frequently included in modern ports (Quinn 1972).

Most ancient coastal efforts were directed to port structures, with the exception of a few places where life depended on coastline protection. Venice and its lagoon is one such case. Here, sea defenses (hydraulic and military) were necessary for the survival of the narrow coastal strips, and impressive shore protection works built by the Venetians are still admired. Very few written reports on the ancient design and construction of coastal structures have survived. A classic treatise by Vitruvius (27 B.C.) relating the Roman engineering experience, has survived (Pollio, Rowland, and Howe 1999). Greek and Latin literature by Herodotus, Josephs, Suetonius, Pliny, Appian, Polibus, Strabo, and others provide limited descriptions of the ancient coastal works. They show the ancients' ability to understand and handle various complex physical phenomena with limited empirical data and simple computational tools. They understood such phenomena as the Mediterranean currents and wind patterns and the wind-wave cause-effect link. The Romans are credited with first introducing wind roses (Franco 1996).

I-3-2. Pre-Roman Times

Most early harbors were natural anchorages in favorable geographical conditions such as sheltered bays behind capes or peninsulas, behind coastal islands, at river mouths, inside lagoons, or in deep coves. Short breakwaters were eventually added to supplement the natural protection. The harbors, used for refuge, unloading of goods, and access to fresh water, were closely spaced to accommodate the safe day-to-day transfer of the shallow draft wooden vessels which sailed coastwise at speeds of only 3-5 knots.

Ancient ports can be divided into three groups according to their structural patterns and the development of engineering skill (Frost 1963).

    a. The earliest were rock cut, in that natural features like offshore reefs were adapted to give shelter to craft riding at anchor.

b. In the second group, vertical walls were built on convenient shallows to serve as breakwaters and moles. Harbors of this type were in protected bays, and often the walls connected with the defenses of a walled town (for example, ancient Tyre on the Lebanese coast). Often these basins were closeable to traffic using chains to prevent the entry of enemy ships (Franco 1996).

c. The third group were harbors that were imposed on even unpromising coasts by use of Roman innovations such as the arch and improved hydraulic cement. Projects like this required the engineering, construction, and financing resources of a major empire.

All ancient ports had one thing in common: they had to be kept clear of silt at a time when mechanical dredging was unknown. This was accomplished by various means. One was by designing the outer parts of the harbor so that they deflected silt-bearing currents. The second was by allowing a controlled current to flow through the port or by flushing it out when necessary by means of channels. For example, at Sidon, a series of tanks (like swimming pools) were cut into the harbor side of a natural rock reef. The tanks filled with clear water that was held in place with sluice gates. When the gates were opened, currents of clear water would flush the inner harbor. Documentary and archaeological evidence show that both Tyre and Sidon were flourishing and powerful ports from the Bronze Age through the Roman era and must therefore have been kept clear of silt for over a thousand years (Frost 1963). Another method of preventing silt consisted of diverting rivers through canals so that during part or much of the year, the flow would enter the sea at location well away from the harbor.

The origins of breakwaters are unknown. The ancient Egyptians built boat basins with breakwaters on the Nile River at Zoser's (Djoser) step pyramid (ca. 2500 B.C.) (Inman 2001). The Minoans constructed a breakwater at Nirou Khani on Crete long before the explosion of Santorini (Thera) in ca. 1500 B.C. The breakwater was small and constructed of material taken from nearby dune rock quarries (Inman 1974, Figure 4). In the Mediterranean, size and sophistication of breakwaters increased over time as the Egyptian, Phoenician, Greco-Macedonian, and Roman civilizations developed and evolved. Breakwaters were built in China but generally at a later date than in the Mediterranean.

Probably the most sophisticated man-made harbor of this era was the first harbor of Alexandria, Egypt, built west of Pharos Island about 1800 B.C. by the Minoans. The main basin, built to accommodate 400 ships about 35 m in length, was 2,300 m long, 300 m wide and 6-10 m deep. Large stone blocks were used in the many breakwaters and docks in the harbor. Alexander the Great and his Greek successors rebuilt the harbor (300-100 B.C.) in monumental scale. The Island of Pharos was joined to the mainland by a 1.5 km breakwater with two openings dividing two basins with an area of 368 hectares (910 acres) and 15 km of quay front. Alexandria is probably best known for the 130m-high lighthouse tower used to guide ships on a featureless coast to the port from 50 km at sea. The multi-storied building was built with solid blocks of stone cemented together with melted lead and lined with white stone slabs. Considered one of the Wonders of the Ancient World, it eventually collapsed due to earthquakes between 1326 and 1349 (Franco 1996, Empereur 1997).

Another feature of the Greek harbors was the use of colossal statues to mark the entrances. Colossal statues of King Ptolemy, which stood at the base of the lighthouse, have been found with the lighthouse debris. Historians report the most famous harbor statue was the 30 m high Colossus of Rhodes, which stood on the breakwater heads. Three ancient windmill towers are still surviving upon the Rhodes breakwater (Franco 1996). Frost (1963) notes that the Greeks had used hydraulic cement long before the Romans.

I-3-3. Roman Times

The Romans introduced many revolutionary innovations in harbor design. They learned to build walls underwater and constructed solid breakwaters to protect exposed harbors. They used metal joints and clamps to fasten neighboring blocks together and are often credited with discovering hydraulic cement made with pozzolanic ash obtained from the volcanic region near Naples, which hardens underwater. Frost (1963) notes that the Greeks had used hydraulic cement long before the Romans. The Romans replaced many of the Greek rubble mound breakwaters with vertical and composite concrete walls. These monolithic coastal structures could be built rapidly and required little maintenance. In some cases wave reflection may have been used to prevent silting. In most cases, rubble or large stone slabs were placed in front of the walls to protect against toe scour. The Romans developed cranes and pile drivers and used them extensively in their construction. This technology also led them to develop dredges. Another advanced technique used for deepwater applications was the watertight floating cellular caisson, precursor of the modern day monolithic breakwater. They also used low, water-surface breakwaters to trip the waves before they reached the main breakwater. The peculiar feature of the vertical wall breakwater at Thapsus (Rass Dimas, Tunisia) was the presence of vents through the wall to reduce wave impact forces. This idea is used today in the construction of perforated caisson breakwaters (Franco 1996).

Using some of these techniques, the Romans built sophisticated breakwaters at Aquileia, Italy (ca. 180 B.C.), and at Caesarea, Israel (ca. 20 B.C.). The southwestern breakwater at Caesarea contained a "forebreakwater" that acted as a submerged reef that "trips" the wave causing it to break and dissipate energy before encountering the main breakwater (Inman 2001).

The largest manmade harbor complex was the imperial port of Rome; the maritime town at the mouth of the Tiber River was named Portus (The Port). It is now some four km from the sea, partly buried under Rome-Fiumicino airport. Despite its importance to the capital of the empire, (300,000 tons/year of wheat from Egypt and France), the harbor always suffered siltation from the river. Trajan, who also built the ports of Terracina and Centumcellae, built Portus' inner hexagonal basin. The port of Centumcellae was built just to serve his villa at a site with favorable rocky morphology. A grandiose engineering project between 107-106 B.C. created a sheltered bathing and boating retreat. Slaves from all parts of the empire excavated a harbor and hauled in massive stones to create an artificial harbor to dampen the force of the waves. After the decline of Portus, it became, and remains, the Port of Rome. After remaining unchanged for over 1,000 years, the inner Roman Basin, which was dredged from rock (200,000 m^3 or 260,000 yd^3), is still in use. Roman engineers also constructed harbors in northern Europe along the main waterways of the Rhine and Danube and in Lake Geneva. They became the first dredgers in the Netherlands to maintain the harbor at Velsen. Silting problems here were solved when the previously sealed solid piers were replaced with new "open"-piled jetties. In general, the Romans spread their technology throughout the western world. Their harbors became independent infrastructures, with their own buildings and storage sheds as opposed to the pre-Roman fortified city-enclosed harbors. They developed and properly used a variety of design concepts and construction techniques at different coastal cites to suit the local hydraulic and morphological conditions and available materials (Franco 1996).

The Romans also introduced to the world the concept of the holiday at the coast. The ingredients for beach holidays were in place: high population density coupled with a relatively high standard of living, a well-established economic and social elite, and a superb infrastructure of roads. From the end of the republic to the middle of the second century of the empire, resorts thrived along the shores of Latium and Capania, and an unbroken string of villas extended along the coast from the seashore near Rome to the white cliffs of Terracina. Fine roads connected these resorts to the capital, allowing both the upper crust and the masses to descend from sultry and vapor-ridden Rome to the sea. For five hundred years, the sybaritic town of Baiae reigned as the greatest fashionable beach resort of the ancient world. Seneca the Younger called Baiae a "vortex of luxury and a harbor of vice," an alluring combination that Romans found irresistible (Lenček and Bosker 1998).

I-3-4. Modern Age

After the fall of the Western Roman Empire, a long hiatus in coastal technology and engineering prevailed throughout most of the European world with a few exceptions. Little is recorded on civil engineering achievements during the Dark and Middle Ages. The threat of attack from the sea caused many coastal towns and their harbors to be abandoned. Many harbors were lost due to natural causes such as rapid silting, shoreline advance or retreat, etc. The Venice lagoon was one of the few populated coastal areas with continuous prosperity and development where written reports document the evolution of coastal protection works, ranging from the use of wicker faggots to reinforce the dunes to timber piles and stones, often combined in a sort of crib work. Protection from the sea was so vital to the Venetians, that laws from 1282 to 1339 did not allow anyone to cut or burn trees from coastal woods, pick out mussels from the rock revetments, let cattle upon the dikes, remove sand or vegetation from the beaches or dunes, or export materials used for shore protection (Franco 1996).

In England, coastal engineering works date back to the Romans, who recognized the danger of floods and sea inundation of low-lying lands. On the Medway, for example, embankments built by the Romans as sea defense remained in use until the 18th century (Palmer and Tritton Limited 1996). The low-lying lands, consisting of recently-deposited alluvial material, were exceeding fertile but were also vulnerable to flooding from both runoff and storm surges. In the Middle Ages, the Church became instrumental in reclaiming and protecting many marshes, and monks reclaimed portions of the Fylde and Humber estuaries. In 1225, Henry III constituted by Charter a body of persons to deal with the question of drainage (Keay 1942).

Across the North Sea, the Friesland area of the Netherlands had a large and wealthy population in the period 500 - 1000 A. D., and increasing need for agricultural land led to building of sea dikes to reclaim land that previously was used for grazing (Bijker 1996). Water boards developed in response to the need for a mutual means to coordinate and enforce dike maintenance. These boards represent some of the earliest democratic institutions in the Netherlands.

Engineering and scientific skills remained alive in the east, in Byzantium, where the Eastern Roman empire survived for six hundred years while Western Rome decayed. Of necessity, Byzantium had become a sea power, sending forth fleets of merchant ships and multi-oared dromonds (swift war vessels) throughout the Black Sea and Mediterranean. When the weary soldiers of the first crusades reached Byzantium's capital city, Constantinople, in 1097, they were amazed and awed by its magnificence, sophistication, and scientific achievements. Constantinople was built on the hills overlooking the Golden Horn, a natural bay extending north of the Bosporus. Marble docks lined the waterfront, over which passed the spices, furs, timber, grain, and the treasures of an empire. A great chain could be pulled across the mouth of the Golden Horn to prevent intrusion by enemy fleets. A series of watch towers extended along the length of the Sea of Marmara, the Bosphoros, and the south shore of the Black Sea, and the approach of an enemy fleet could be signaled to the emperor within hours by an ingenious code using mirrors by day and signal fires by night (Lamb 1930).

The Renaissance era (about XV - XVI centuries) was a period of scientific and technologic reawakening, including the field of coastal engineering. While the standards for design and construction remained those developed primarily by the Romans, a great leap in technology was achieved through the development of mechanical equipment and the birth of the hydraulic sciences including maritime hydraulics (Franco 1996). "The Italian School of Hydraulics was the first to be formed and the only one to exist before the middle of the 17th century" (Rouse and Ince 1963). Leonardo da Vinci (1465-1519), with his well-known experimental method, based on the systematic observation of natural phenomenon supported by intellectual reasoning and a creative intuition, could be considered the precursor of hydrodynamics, offering ideas and solutions often more than three centuries ahead of their common acceptance. Some of his descriptions of water movement are qualitative, but often so correct, that some of his drawings could be usefully included in a modern coastal hydrodynamics text. The quantitative definition and mathematical formulation of the results were far beyond the scientific capabilities of the era. Even so, da Vinci was probably the first to describe and test several experimental techniques now employed in most modern hydraulic laboratories. To visualize the flow field, he used suspended particles and dyes, glass-walled tanks, and movable bed models, both in water and in air. The movement from kinematics to dynamics proved impossible until the correct theory of gravitation was developed, some two centuries latter by Sir Issac Newton (Fasso 1987). The variety of hydro kinematics problems dealt with in da Vinci's notebooks is so vast that it is not possible to enumerate them all in this brief review. In the 36 folios (sheets) of the Codex Leicester (1510), he describes most phenomena related to maritime hydraulics. Richter (1970) provides an English translation of da Vinci's notebooks (Franco 1996). The scientific ideas of the Italian Renaissance soon moved beyond the confines of that country, to the European countries north of the Alps.

I-3-5. Military and Civil Engineer Era

After the Renaissance, although great strides were made in the general scientific arena, little improvement was made beyond the Roman approach to harbor construction. Ships became more sea-worthy and global navigation became more common. With global navigation came the European discovery of the Americas, Australia, New Zealand, Indonesia, and other areas of the world, soon followed by migration and colonization. Trade developed with previously unreachable countries and new colonies. France developed as the leader in scientific knowledge. The French "G'enie" officers, who, along with their military task, were also entrusted with civilian public works, are reportedly the direct ancestors of modern civil engineers. S'ebastien le Prestre de Vauban (1633-1707) was a builder of numerous fortresses and perfected the system of polygonal and star shaped fortifications. His most eminent public works project was the conversion of Dunkirk into an impregnable coastal fortress. Apart from the construction of several forts, there were extensive harbor and coastal works, including the excavation of canals and harbor basins, the construction of two long jetties flanking the entrance channel, and the erection of storehouses and workshops. A great lock, a masterpiece of civil engineering, was built at the entrance to the Inner Harbor. Vauban himself designed and supervised the lock construction. Unfortunately, no more than 30 years after its completion, the fortress was destroyed as a consequence of the Spanish War of Succession. Vauban's projects provide a good example of engineering methods and lucidity. They consisted of an explanatory memorandum, several drawings, and a covering letter. The memorandum had four sections: (1) general background of the scheme; (2) detailed descriptions of the different parts, with references to the drawings; (3) cost estimates; (4) features and advantages of the work. It was during this time that the term "Ingenieur" was first used in France, as a professional title for a scientifically-trained technician in public service.

While France enjoyed a leading position in Europe with regard to exact sciences and their applications to technical problems, a social and economic revolution later known as the "Industrial Revolution" was taking place in England. The riding-horse and the packhorse gave way to the coach, the wagon and the barge. Hard roads and canals replaced the centuries old soft roads and trails, dusty in dry weather and mud-bound during rains (Straub 1964). Steam power allowed industry to be concentrated in factories that required continuous supply of raw materials and export of manufactured goods.

In the 18th and 19th centuries, advances in navigation and mathematics, the advent of the steam engine, the search for new lands and trade routes, the expansion of the British Empire through her colonies, and other influences, all contributed to the revitalization of sea trade and a renewed interest in port works. As the volume of shipping grew, more vessels were needed and as the dimensions of the new vessels became larger, increased port facilities were necessary. Ports of the world experienced growing pains for the first time since the Roman era, and, except for the interruption caused by two world wars, port needs continue to grow (Quinn 1972).

I-3-6. United States Army Corps of Engineers

Since the formation of the United States, Army engineers and the Corps of Engineers have been responsible for or intimately associated with a wide variety of civil projects and improvements to waterways, ports, and navigation systems. The following paragraphs summarize the history of the U.S. Army Corps of Engineers (USACE) and outline some of the Corps' early efforts in coastal and navigation improvements.

The origins of the USACE date to June of 1775, at the beginning of the American War of Independence, when the Second Continental Congress authorized General Washington to assign a "chief engineer" for the "grand army" (Parkman 1978). General Washington selected Colonel Richard Gridley, a seasoned artilleryman, who had been preparing a line of fortifications around Boston during the early weeks of the war. Military operations during the war underscored the need for an efficient body of engineers, and in March of 1779, the Continental Congress finally authorized a separate and distinct "corps of engineers," to be commanded by Louis LeBègue Du Portail, an officer recruited by the American mission in France. The corps was a vital unit of the Continental Army until disbanded in November 1783 with the arrival of peace. When war between France and England broke out in the 1790s Congress authorized President Washington to begin construction of a system of fortifications along the coast. In 1802, in anticipation of the European belligerents signing a treaty of peace, Congress cut back and reorganized the army and created a separate corps of engineers, limited at that time to sixteen officers (Parkman 1978). The Act of March 16, 1802 had other far-reaching consequences, as it provided further that the Corps was to constitute the personnel of a military academy at West Point. Congress had recognized the almost complete absence of trained military and civil engineers in the United States, and, in effect, established a national college of engineering. West Point was the only school in the country to graduate engineers until 1824, when Rensselaer Polytechnic Institute was formed. Quickly becoming the growing nation's primary source of engineering expertise, the Corps first concentrated on constructing and maintaining strategically-placed coastal fortifications to repel naval attacks. But soon it became concerned with civil functions as it planned and executed the national internal improvement program initiated in the 1820s (Maass 1951).

Until the early 1800s, little maintenance or improvement was done to harbors or rivers, and maintaining navigability of waterways was considered the responsibility of the states or private interests. What little the Federal government had done was carried out by the Treasury Department, which had assisted navigation by erecting lighthouses, beacons, buoys, and public piers. In 1818, John C. Calhoun, then Secretary of War, recommended that the Corps of Engineers be directed to improve waterways navigation and other transportation systems because these civil works would facilitate the movement of the Army and its materials while contributing to national economic development (USACE 1978). Congress accepted Calhoun's recommendations and passed the landmark General Survey Act, which President James Monroe signed into law on April 30, 1824. It directed the President to use Army engineers to survey roads and canals. By the mid-1820s Corps of Engineers officers were busy surveying the Ohio and lower Mississippi Rivers and the Great Lakes, identifying navigation impediments, and proposing improvements and new routes.

Only a month later, on May 24, 1824, President Monroe signed the first Rivers and Harbors Act, which authorized the President to appropriate Federal monies to improve navigation on the Ohio and Mississippi Rivers. By 1829, Army engineers were using steam-powered snagboats to remove snags and floating trees and to dig out sandbars that impeded river traffic. Subsequent acts authorized a wide variety of internal improvements and assigned Army engineers to direct and manage these projects. Work soon began on a number of challenging locations that were deemed critical for the growth of a growing nation.

Hazardous navigation conditions on the Great Lakes also called for the rapid improvement of harbors. With the passage of the Rivers and Harbors Act, Congress voted a $20,000 appropriation for deepening the channel at the harbor of Presque Isle (at Erie, Pennsylvania) on Lake Erie (Drescher 1982). Signaling the beginning of federal involvement in the development of harbors on the Great Lakes, the USACE now maintains over 600 navigation projects throughout these waterways.

One of the USACE's first civil projects on an ocean coast was repairing Long Beach at Plymouth, Massachusetts. The beach was a long, narrow sand spit that formed the town's harbor. Constantly endangered by waves and wind, it had been a subject of concern to the citizens of the town as early as 1702, when they made it a crime to fell its trees or fire the grass. The congressional appropriation of $20,000 on May 26, 1824, "to repair Plymouth Beach in the state of Massachusetts, and thereby prevent the harbour at that place from being destroyed" initiated the Corps' civil works mission in New England. Corps officers supervised local agents, who built a cribwork breakwater along the beach's outer shore and erected brush fences and planted grass to stabilize the sand. Similar projects were undertaken at nearby Duxbury and other beaches in New England (Parkman 1978). The pattern whereby Corps specialists supervise local contractors has continued to this day for most civil works projects.

Over the succeeding century and a half, the USACE's role in civil works grew dramatically, in step with the growth of the nation's population and economy. To adequately cover this interesting story in the CEM would risk doubling its size, so readers are referred to a series of books that document the history of each district (Table I-3-1).

Presently, USACE officers and a large contingent of non-military government employees maintain a navigation system of more than 40,000 km (25,000 miles) and 219 locks and dams connecting large regions of the country. Of the nation's top 50 ports active in foreign waterborne commerce, over 90 percent require regular dredging (Waterborne Commerce Statistics Center 1999). Over 300 million cubic meters of dredged material are removed from navigation channels each year. In 1997, the USACE contracted for the dredging of about half this total (157 million m^3, see Figure I-3-1) at coastal sites only. This does not include inland waterways (Hillyer 1996).

Most inlets and harbors used for commercial navigation in the United States are protected and stabilized by hard structures. The USACE built most of the structures and is responsible for maintaining even a larger number since the Federal Government assumed responsibility for some state and local projects. Figure I-3-2 summarizes the locations of the Federal projects (Hillyer 1996).

Many U.S. coastal urban and recreational centers are protected by erosion control and storm damage reduction projects constructed cooperatively by the USACE, state, and local governments (Figure I-3-3). Although most of the 83 Congressionally authorized shore protection projects are in densely developed areas, some were constructed primarily for recreation and are associated with public or park beaches (Sudar et al. 1995).

Table I-3-1
Selected Bibliography of USACE History

Region	Reference
General history of USACE	USACE 1998
Alaska District	Jacobs 1976
Atchafalaya Basin	Reuss 1998
Baltimore District	Kanarek 1978
Buffalo District	Drescher 1982
Charleston District	Moore 1981
Chicago District	Larson 1979
Detroit District	Larson 1981
Galveston District	Alperin 1977
Honolulu District	van Hoften 1970
Jacksonville District	Buker 1981
Little Rock District	Rathbun 1990
Los Angeles District	Turhollow 1975
Mobile District	Davis 1975
New England Division	Parkman 1986
New Orleans District	Cowdrey 1977
New York District	Klawonn 1977
North Atlantic Division	Chambers 1980
Ohio River Division	Johnson 1992
Pacific Ocean Division	Thompson 1981
Philadelphia District	Snyder and Guss 1974
Pittsburgh District	Johnson 1979
Portland District	Willingham 1983
San Francisco District	Hagwood 1982
Savannah District	Barber 1989
Seattle District	Seattle District 1969
St. Paul District	Merritt 1979
St. Lawrence Seaway	Becker 1984
St. Louis District	Dobney 1978
Wilmington District	Hartzer 1984

History of Coastal Engineering

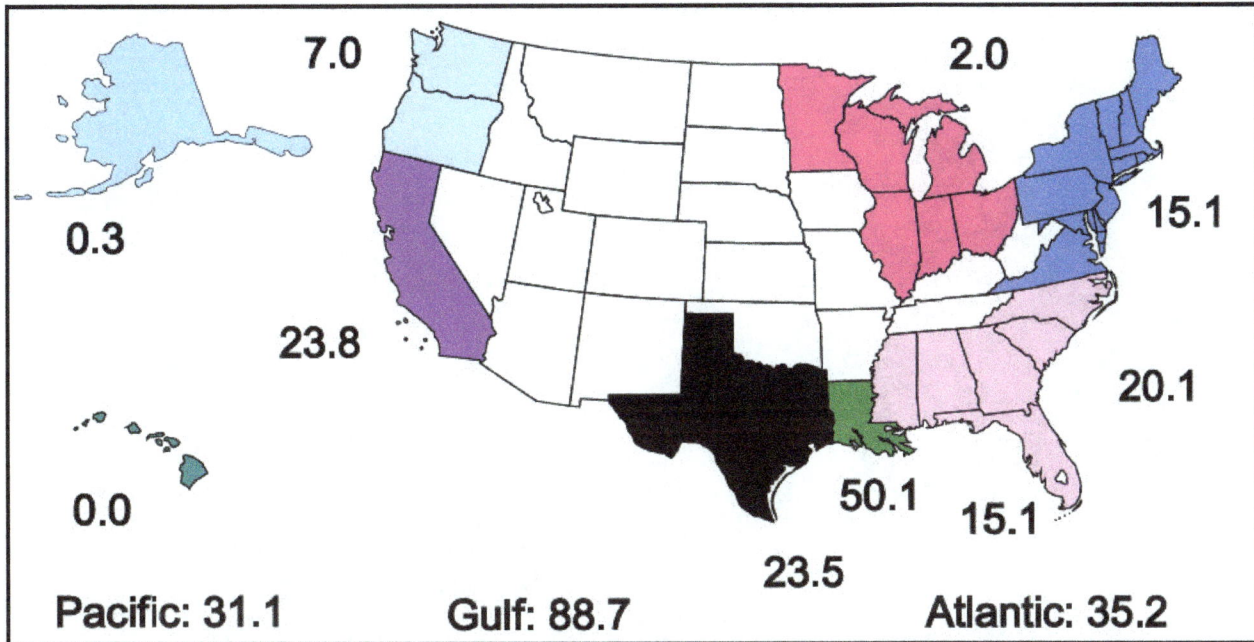

Figure I-3-1. Fiscal year 1997 dredging by the U.S. Army Corps of Engineers at coastal projects (million m^3). These totals do not include dredging of inland waterways and rivers, but do include Great Lakes ports (from Hillyer 1996)

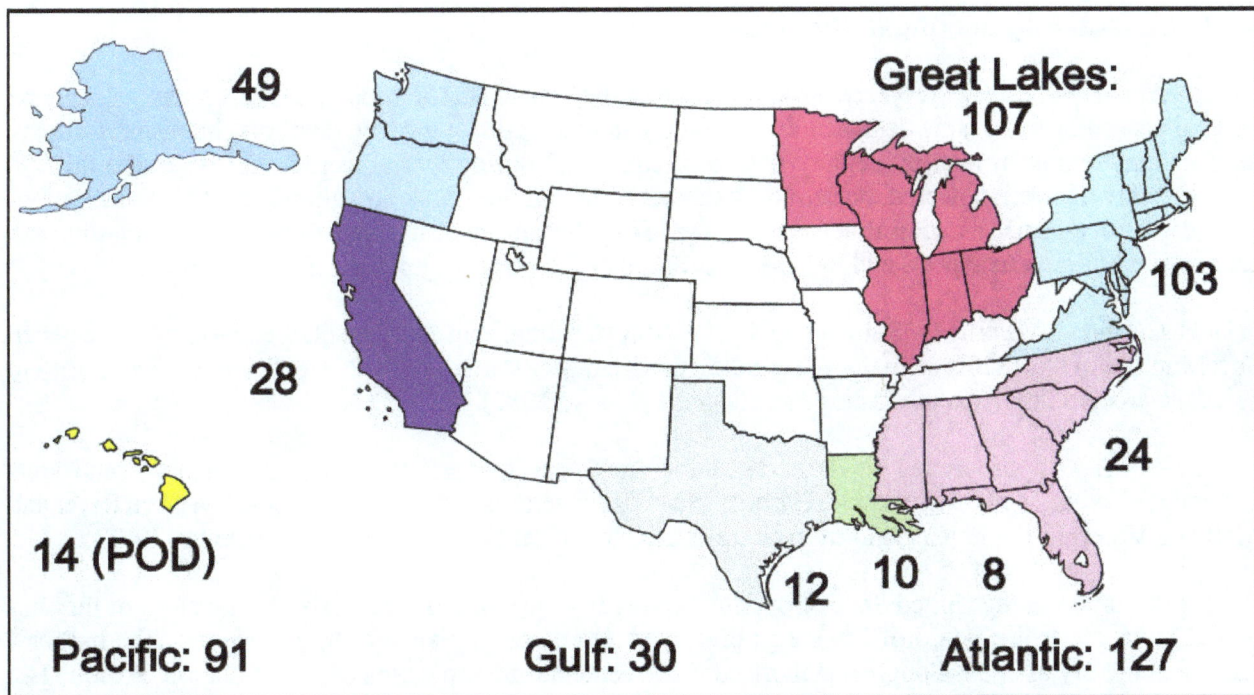

Figure I-3-2. Federally maintained deep-draft and small boat harbors with structures (from Hillyer 1996)

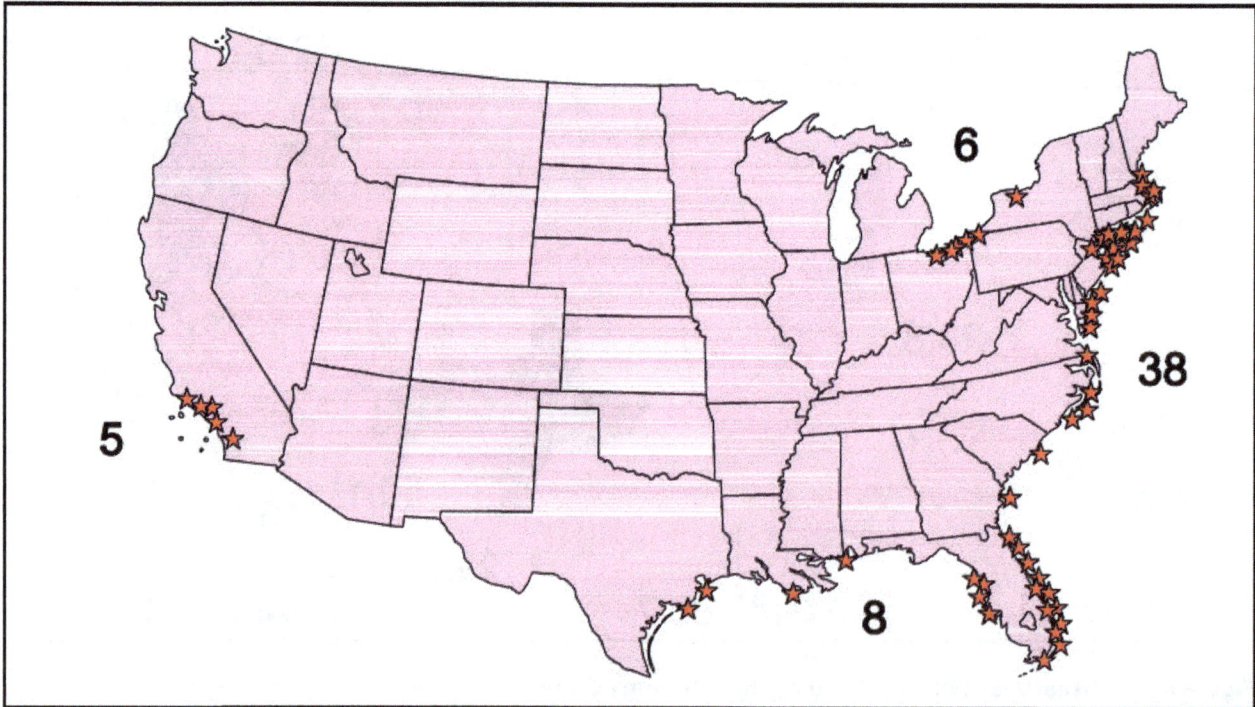

Figure I-3-3. Location of 57 large Congressionally authorized shore protection and beach erosion control projects. Some 26 small projects with limited scope and low cost (less than a few hundred thousand dollars) are not shown (from Hillyer 1996)

I-3-7. Coastal Engineering in the United States[1]

a. Nineteenth century projects. From the birth of the United States through the 18th century, local and Federal coastal projects were designed to accommodate and facilitate growth. Harbors that were usable in their natural setting in the 18th century had to be improved during the industrial age to service the 19th century's larger, steam powered, ocean-going vessels. Though the Corps' attention focused on navigation-related improvements, its coastal activities ranged from beach reconstruction to blasting better shipping channels to building new ports and lighthouses. Some of the early projects are summarized below:

(1) Currents sweeping by Sullivan's Island, South Carolina, caused substantial erosion that threatened Fort Moultrie in the 1820s. A major reclamation program was started by the USACE under its authority to solve the erosion problems of existing fortifications (Moore 1982).

(2) Congress appropriated funds in 1826 for the Corps to combat erosion of valuable sand spits protecting Duxbury and Provincetown Harbors in Massachusetts and to construct jetties at Warren River and Martha's Vineyard to prevent sand from being carried by currents into the harbors (Parkman 1978).

(3) Attention of the United States Government was first directed to Erie Harbor at the close of the War of 1812. In 1823, the Board of Engineers presented an elaborate plan for improvement of the harbor's entrance. In May of 1824, Congress authorized improvements and protection of the vulnerable Presque Isle Peninsula (Goreki and Pope 1993). Engineering work continues at Presque Isle to this day (Figure V-3-10).

---

[1] One of the most detailed discussions of the history of coastal engineering in the United States is Weigel and Saville 1996.

(4) In 1830, Army Engineers surveyed and made recommendations for the improvement of Baltimore Harbor, Maryland. A prolonged program of channel improvement began in 1852, and by the summer of 1872, as many as 13 dredges were engaged in the excavation of the waterway. By the time of the Spanish-American war in 1898, Baltimore had become one of the world's major ports (Kanarek 1976 pp. 41-59).

(5) Buffalo, New York, and Cleveland, Ohio, grew from frontier villages to manufacturing and commercial centers in a little over a century because of their locations at the terminus of water and rail routes connecting the grain-rich areas of the west to the eastern urban centers. The economic lives of the two cities depended on the construction and maintenance of harbor facilities such as seawalls, jetties, breakwaters, and dredged channels. As a result of successive harbor improvement projects, they have become major cities on the Great Lakes (Drescher 1982). Much of the 19th century development of the mid-West and the Great Lakes occurred as European immigrants traveled through the port of New York, along the Erie Barge Canal through Buffalo, and on to points further west.

(6) Hell Gate, a one-mile section of the East River that connected Long Island Sound with New York Harbor, had very strong currents that sliced around rocks and islands and ran back and forth because the tides in the harbor and sound did not coincide. In 1845, New York city began an effort to open the East River to navigation and in 1852, the Corps tackled the immensely difficult task of developing new technology for underwater excavation and blasting that would be required to clear Hell Gate for navigation. The project was completed 30 years later (Klawonn 1977 pp. 69-93).

(7) In 1868, Congress responded to request for assistance from California that resulted in a long productive period of Federal, State, and local cooperation. The development of the California coast with rail connections to modern, deepwater ports at San Diego, Los Angeles, San Francisco, and Oakland was the ultimate result (Turhollow 1975 pp. 20-48).

(8) Following the devastating 1900 hurricane at Galveston, Texas, which drowned over 6,000 of its citizens, the city assigned three civilian engineers the task of developing the safest and most efficient means to protect the city from similar future floods. Based on their study, the city constructed a 5,360-meter (17,600-foot) curved face concrete seawall, (Figure V-3-5). The city was elevated several meters using sand pumped from Galveston Bay onto the beach behind the seawall. At the same time, Congress authorized a connecting seawall to protect the port and military reservation at Fort Crockett. The 4,900-ft extension was constructed from 1904 to 1905 (Alperin 1977, pp. 237-244).

b. *Nineteenth century coastal engineering.* In 19th century United States, most engineering in the coastal area consisted of the application of principles well known to engineers accustomed to dealing with rivers. There was little concern about the unique nature of the coast, and studies of the effects of wind and waves upon the shore were sporadic, desultory, and unscientific. Trial and error, frequently accompanied by innovation was the teaching tool of the day. Improvement of the St. Johns River mouth at Jacksonville, Florida provides a good example. A continuously shifting sinuous channel through the bar made navigation difficult, so in the 1850s, a citizens group petitioned the USACE for help in dealing with the sandbar problem. One solution proposed was to put the scouring power of the current to work by constructing jetties. The USACE engineers preferred to try clearing the channel by frequent dredging and raking during the strongest phase of the ebb. These attempts failed, and in 1878 influential citizens hired Captain James B. Eads to study the problem. The 1878 Eads report recommended constructing two converging jetties to create a stable deep channel out to sea. His report contained principles of seacoast engineering, sketches of the tidal prism, and estimates of the area that could be maintained. The sophisticated technology to confirm Eads' findings would not be available well into the 20th century. Largely as a result of Eads' success constructing the jetties at the mouth of the Mississippi River, the USACE adopted a modified version of his jetty plan for improving the St. Johns River entrance (Buker 1981, pp. 69-82) (Figure I-3-4).

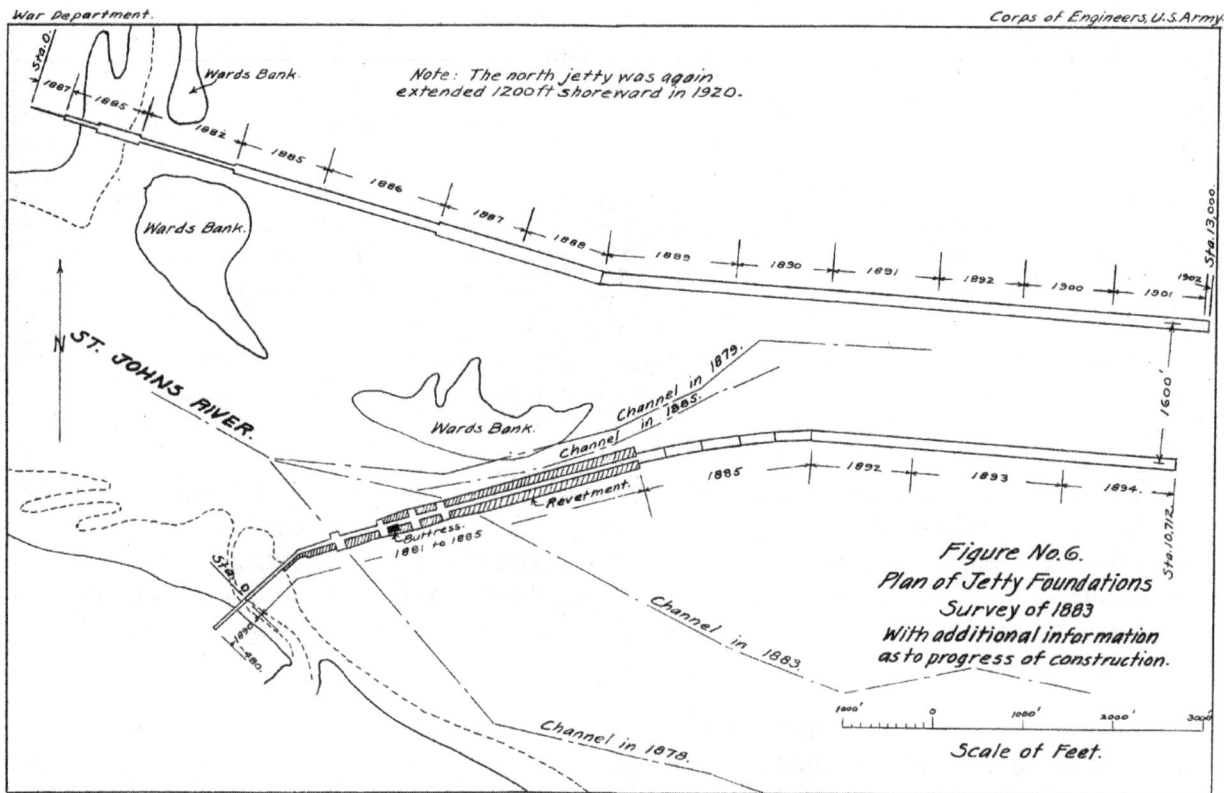

Figure I-3-4. Converging jetty system designed by James Buchanan Eads for the mouth of the St. Johns River, Florida. Figure courtesy of USAED, Jacksonville

Colonel Quincy Gillmore, familiar with Eads' plan for Jacksonville, used a similar plan to construct the jetties at Charleston Harbor, South Carolina, between 1878 and 1893. From the barrier islands on each side of the harbor entrance, the USACE constructed two converging jetties followed by a parallel section. The near shore portion of the jetties was constructed to just below the low tide water surface, thereby serving as a weir and allowing the flood tide to come in normally. During the ebb tide, the bottom currents were channeled through the parallel section (constructed higher, with the seaward quarter above high tide) toward the bar and this scouring action kept the new channel clear (Moore 1982). A similar plan was used at the mouth of the St. Mary's River at the Florida-Georgia border. Major George L. Gillespie, District Engineer in Portland, OR, submitted a plan for a dike at the mouth of the Columbia River to concentrate river currents and tides to scour a deep channel. Construction commenced on the south jetty in 1884, a project that had to overcome fierce winter storms and hazardous working conditions (Willingham 1983).

c. Early coastal development and shore protection. From the early days of settlement to the present, Americans have built in the coastal area. During the 1600 and 1700s, the original colonies owed their prosperity largely to the availability of good natural harbors, rich nearby fishing grounds, and active trade with the Caribbean and Europe. As the giant continent was explored and settled in the 1800s, rivers and the Great Lakes became the prime mode of moving goods and people to and from distant towns. In the 20th century, a new social phenomena arose that resulted in an ever-increasing interest in the coastal zone: more and more Americans achieved the economic means and leisure time to enjoy the beach for recreational purposes. Even before beaches became popular vacation destinations, engineers constructed structures to aid navigation, to halt erosion, and to protect shore front development from storm surge. They designed bulkheads, revetments, and seawalls to hold the shoreline in place. Generally, these designs were successful, with Galveston, Texas, and San Francisco, California, being two examples of early seawall construction. Other structures, such as groins and jetties, impeded longshore transport of sand. Groins are intended to protect a finite beach section, while jetties keep sand out of the navigation channel between the jetties, define and maintains the harbor entrance channel, and provide calm water access to the harbor facilities (Figures I-3-5, I-3-6). For jetties built along uninhabited coastal areas in the 19th and early 20th centuries, the buildup of sand on the up-drift beach and the loss of sand on the downdrift beach was considered a minor consequence compared to the major benefits of ocean navigation trade (Figure I-3-7). In nearly every instance, these harbor structures interrupted the alongshore movement of sand and starved nearby downdrift beaches (USACE 1971), but it was not until the shore was developed in the later 20th century that this interruption of sand transport was regarded as a problem.

d. Early 20th century beach development and the Engineering Advisory Board on Coastal Erosion. As urbanization and congestion increased, the more affluent escaped to the seashore, where resorts arose to accommodate them. Until the age of the automobile, these resorts remained small isolated coastal enclaves tied to the hinterland by rails. The technical revolution brought electric trains, automobiles, gasoline-powered pleasure boats, labor-saving devices for the home, and a new era of leisure to a prospering nation (USACE 1971; Morison and Commager 1962). Electricity provided convenient power to energy-poor barriers. Changing morals allowed people to sunbathe and enjoy the hedonism of the beach experience. And with the growing use of the automobile, beach-goers in increasing numbers followed newly-built roads to the coast. Concern about shore erosion grew as more people acquired property and built homes and businesses, assuming a stable shoreline.

The New Jersey shore, close to the New York and Philadelphia urban areas, was one of the first highly developed shorelines (Figures I-3-8 and I-3-9). During the period 1915 to 1921, three hurricanes and four tropical storms battered the Jersey shore, causing severe beach erosion. In New Jersey, millions of dollars were spent on uncoordinated and sometimes totally inappropriate erosion control structures which often produced results that were only minimally effective, and, in some cases, counterproductive (Hillyer 1996). Engineers and city managers soon realized that individual property owners were incapable of dealing with coastal erosion and that a broader approach was necessary. In 1922, because of rapidly eroding shorelines and revenue losses to the coastal communities, the State funded and appointed an Engineering Advisory Board on Coastal Erosion. Its only recommendation was that further research was needed (Moore and Moore 1983).

In contrast to the haphazard development of the Jersey shore, some of the early large-scale coastal projects proved to be remarkably successful social and engineering accomplishments. America's first large engineered beach fill was the boardwalk and recreational beach on Coney Island in 1922 - 1923 (Farley 1923). With the completion of the project, immigrants and factory workers could escape the sweatshops of the sweltering city and enjoy a (crowded) Sunday at the beach for only a nickel subway ride (Figure I-3-10; Stanton 1999). This was followed by the ambitious construction of the Jones Beach Parkway by Robert Moses and the Long Island State Park Commission in 1926 - 1929, during which more than 30 million m^3 of sand were pumped to create Jones Island (DeWan 1999; Kana 1999) (Figure I-3-11). In Chicago, the entire waterfront was reshaped between 1920 and 1940 with the addition of over 14.2 square km of fill, resulting in one of America's premier urban parks (Chrzastowski 1999). These were city- and state-sponsored projects, with minimal input by the Federal Government.

e. American Shore and Beach Preservation Association. Delegates (85) representing 16 states met at Asbury Park, New Jersey, in 1926, to discuss their growing coastal zone problems. After the first meeting, two more, following shortly thereafter, led to the formation of the American Shore and Beach Preservation Association (ASBPA). The Association brings together a cross section of engineers, public officials, State and Federal personnel, and coastal property owners. Their aim is that "Man must come to the rescue of the beaches." Members see themselves as leaders and teachers in a conservation movement to fight shore and beach erosion (Patton 1934). Their influence in State and Federal governments and continued interest in coastal zone issues is responsible for many of the laws and actions to protect the U. S. shores and beaches.

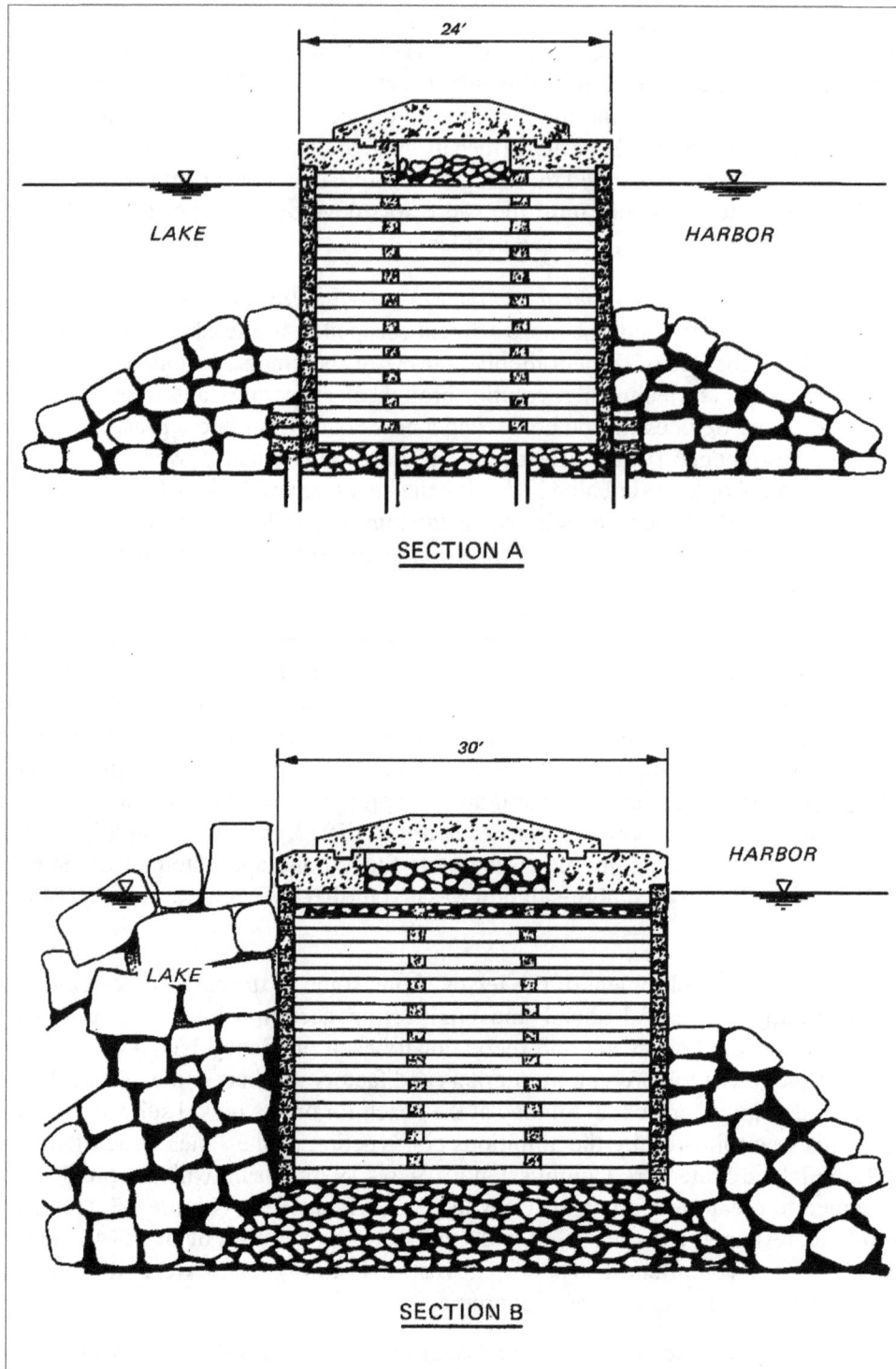

Figure I-3-5. Example of wood crib breakwater typical of the construction technique used in the Great Lakes during the late 1800's, in this case from Calumet Harbor, Illinois. Skilled wood craftsmen built the cribs on land or on a barge, floated them into place, and sank them using rock and gravel fill. Almost 1,000 of these breakwaters were built around the Great Lakes, and many of them, now a century old, need rehabilitation.

Figure I-3-6. Construction of Fire Island jetty on Democrat Point, at the west end of Fire Island, Long Island, New York. Many early USACE jetties were built from the land moving seaward using materials brought in by train. Photograph taken 3 Jan 1940, from Beach Erosion Board Archives.

f. The Board on Sand Movement and Beach Erosion. On January 29, 1929, Major General Edgar Jadwin, Chief of Engineers, USACE, issued a Special Order creating a four-officer board "to investigate and report on the subject of sand movement and beach erosion at such localities as may be designated by the Chief of Engineers." The Chief of Engineers designated Jamaica Bay, New York, and Cold Springs Inlet, New Jersey, as the first two projects for investigation by the "Sand Board" as it became familiarly known. The board employed two consultants, Dr. Douglas W. Johnson of Columbia University and Professor Thorndike Saville of the University of North Carolina. With their advice and assistance, a list of field experiments was prepared. The northern coast of New Jersey, with its area of active erosion and numerous shore protection installations was selected for conducting the experiments. Lieutenant Leland K. Hewitt was assigned to the board in April 1929, to conduct the experiments. Morrough P. O'Brien, recently returned from Freeman Scholarship studies in Europe, was borrowed from the University of California to provide expert assistance in the studies. Hewitt and O'Brien set up their headquarters in Long Branch, New Jersey, and began assembling equipment and personnel to carry out their task. Thus began a USACE research program destined to have far-reaching effects (Wilson and Eaton 1960)

g. The Beach Erosion Board. During this period, the ASBPA, under the leadership of J. Spencer Smith, its persuasive president, was engaged in a campaign to bring combined Federal, State and local effort to bear upon the U. S. beach preservation problems. After two years of Congressional hearings, Section 2 of the River and Harbor Act of 1930, authorized and directed the Chief of Engineers "to cause investigations and studies to be made in cooperation with appropriate agencies of the various states on the Atlantic, Pacific, and Gulf coasts and on the Great Lakes, and Territories, with a view of devising effective means of preventing erosion of the shores of coastal and lake waters by waves and currents . . ." The new law also provided for the creation of the Beach Erosion Board (BEB) by the provision "that there shall be organized under the Chief of Engineers, United States Army, by detail from time to time from the Corps of Engineers and from the engineers of state agencies charged with beach erosion and shore protection, a board of seven members,

Figure I-3-7. Construction of the jetty on the east side of Rockaway Inlet, Long Island, New York, 12 July 1932. The urban area on the opposite side of the inlet is the east end of Coney Island, and Jamaica Bay is in the distance. In seven decades, the fillet has filled with sand to approximately the end of the jetty seen in this image. Photograph from Beach Erosion Board Archives.

of whom four shall be officers of the Corps of Engineers and three shall be selected with regard to their special fitness by the Chief of Engineers from among the state agencies cooperating with the War Department. The board will furnish such technical assistance as may be directed by the Chief of Engineers in the conduct of such studies as may be undertaken and will review the reports of investigation made . . ." Obviously, reconsideration of the mission and need for the original Board on Sand Movement and Beach Erosion was required. Since the new law defined the functions of the BEB as being related to cooperative studies with the states, It was decided to create two boards, one known as the Shore Protection Board (SPB), that would conduct investigations and report upon problems concerning federal property shore protection and the other, the BEB to have similar responsibilities with respect to cooperative studies. Members of the SPB consisted of the military members of the BEB plus the District Engineer for the concerned locality. For the next sixteen years the two boards shared the same staff and headquarters until the SPB was abolished and its duties transferred to the BEB. (Wilson and Eaton 1960)

h. BEB focus on basic research (Willingham 1983). The Corps had, historically, not favored expenditure of Federal funds to protect private property, whether in river basins or coastal flood plains. During the 1930's, however, attitudes began to change, and Congress expanded Federal

Figure I-3-8. Atlantic City, New Jersey, at the mouth of Absecon Inlet, 15 September 1944. This is one of a series of images taken after the hurricane of 14 September, a Category 3 storm that caused 390 deaths in the northeast U.S. This photograph illustrates the degree of urban development along this coast. Photograph taken from a blimp from the U.S. Naval Air Station, Lakehurst, NJ. Official U.S. Navy Photograph (from Beach Erosion Board Archives)

participation in coastal protection. A significant legislative event was the passage in 1936 of an Act wherein it was declared the policy of the United States "... to assist in the construction where Federal interests are involved, but not the maintenance, of works for the improvement and protection of the beaches along the shores of the United States, and to prevent erosion due to the action of waves, tides, and currents, with the purpose of preventing damage to property along the shores of the United States, and promoting and encouraging the healthful recreation of the people..." The Act further authorized the Board "to publish from time to time such useful data and information concerning the protection of beaches as the Board may deem to be of value to the people of the United States..." The act also required that the Board "in making its report on any work or project relating to shore protection shall,

Figure I-3-9. Construction of Manasquan Inlet jetties, New Jersey, October 2, 1930, view looking north. Material for the jetties was supplied via an elevated roadway that extended out to sea from the land. Note that sand is already accumulating on the south (lower) side of the south jetty. The shoreline is continuous, and at this site the inlet was dredged after the jetties were completed. Other man-made openings that are now Federal navigation projects include Panama City Inlet, Florida, Duluth Cut, Minnesota, and Aransas Pass, Texas. (Photograph from Beach Erosion Board archives)

in addition to any other matters upon which it may be required to report, state its opinion as to the advisability of adopting the project; what Federal interest, if any, is involved in the proposed improvement; and what share of the expense, if any, should be borne by the United States." (Cited in Wilson and Eaton 1960).

Although there was substantial support in Congress for federal aid in coastal protection, much difficulty was encountered in determining the proper extent of such aid. The BEB, lacking more specific instructions from Congress, interpreted "federal interest" as pertaining only to the interest of the United States as a landowner of shore property. This resulted in practically no recommendations for federal aid by the BEB during the 1930s. Other Federal agencies were concerned with putting people to work during the depression and interpreted the 1936 Act differently. The Works Progress Administration built revetments, dikes, retaining walls, and jetties on North Carolina's Outer Banks at a cost more than $4 million. The Corps held back on coastal construction projects because of uncertainties about predicting conditions at individual coastal sites revealed by board survey reports. The BEB, driven by professional curiosity, undertook

Coney Island, New York
1941

Figure I-3-10. Coney Island in 1941, on the eve of World War II. For only a 5-cent subway ride, workers from the sweltering city could relax at the beach (from Beach Erosion Board archives)

scientific investigations into coastal processes despite the lack of authorization in either the 1930 or 1936 acts. By the beginning of World War II, the BEB was publishing technical reports and memoranda on its research results (Moore and Moore 1991).

i. Dalecarlia reservation and World War II. Expansion of military activities prior to our entry into World War II required the removal of the Washington Engineering District and the BEB from the Navy Building on Constitution Ave. in Washington, D.C. to a small office building on the Dalecarlia reservation in Washington. The new site was adequate for expansion of facilities, which was to follow, and admirably served the BEB's needs. Upon our entry in the war, need soon arose for intelligence to meet amphibious operations requirements and research to explore means of providing expedient harbor facilities. Logically, the Chief of Engineers called on the BEB staff for assistance on these problems. In addition, the BEB was tasked to train intelligence teams to staff the various military commands. Although these activities were directed toward military requirements, they provided much additional data and knowledge for later use in the Board's peacetime mission. Interest and activity in Congress to clarify the problem of federal aid for shore protection resumed as the war drew to a close. Public Law 166, enacted on July 31, 1945, substituted "public interest" for "Federal interest" as previously used in the 1936 Act, and a year later, Public Law 727 spelled out the conditions and limitations for federal aid for shore protection works. Only publicly owned shores were eligible and the Federal contribution could not exceed one-third of the first cost of protective works with no contribution toward maintenance cost. During the Korean War, most of the Board's staff was again diverted to military efforts. During this lull in preparation of cooperative studies, Technical Report No. 4, *Shore Protection Planning and Design*, was produced. The report was a manual of coastal engineering that summarized the knowledge gained by the BEB and representing its current technical doctrine. Cooperative studies resumed at an accelerated pace with the end of the Korean War. Congress

Figure I-3-11. Jones Island State Park, Long Island, New York, 4 July, 1958. One of the nation's premier urban beaches, Jones Beach is only 45 km from Manhattan, but traffic can be problematic on some days. The legend on the photograph states that there are 11,117 cars in the parking lots. From Beach Erosion Board archives

amended the law to permit Federal aid to privately owned shores when a public benefit resulted and to permit aid toward periodic beach nourishment costs. The 1954 and 1955 River and Harbor Acts authorized sixty-two federal aid projects, a significant increase over the five projects authorized up to that date (Wilson and Eaton 1960).

j. BEB accomplishments. In summary, the BEB made substantial progress toward establishing sound coastal engineering techniques and established a research impetus in coastal processes. It commenced the collection of a permanent record of pertinent data and provided a manual on the best techniques to address specific shore erosion problems. The presence of Murrough P. O'Brien and Thorndike Saville on the Beach Erosion Board for such an extended period provided a continuity of objective and effort unique in public service (Wilson and Eaton 1960). During its 33-year existence, the BEB reviewed 149 cooperative study reports and two Federal surveys of beach erosion problems. The BEB reviewed 114 of the cooperative studies following the 1946 legislation allowing Federal participation in construction cost and recommended 72 as Federal projects. The BEB published 135 technical memoranda and four technical reports, 130 of which were published after the 1946 legislation authorizing the BEB to make general investigations and publish technical reports (Moore and Moore 1991).

Figure I-3-12. Workers planting grass on a beach restoration project. The date and location of this image were not recorded, but the scene is likely the Outer Banks. Many dune and beach restoration efforts, sponsored by the National Park Service and other agencies during the late 1930s, also served as work relief efforts for a nation trying to recover from the Great Depression. Photograph from the Beach Erosion Board Archives

k. Evolution of shore protection and the shift from structures to beach nourishment. Prior to the 1950s, the general practice was to use hard structures to protect against beach erosion or storm damages. These structures were usually coastal armoring such as seawalls and revetments or sand-trapping structures such as groins. During the 1920s and '30s, private or local community interests protected many areas of the shore using these techniques in a rather ad hoc manner. In certain resort areas, structures had proliferated to such an extent that the protection actually impeded the recreational use of the beaches. Erosion of the sand continued, but the fixed back-beach line remained, resulting in a loss of beach area. The obtrusiveness and cost of these structures led the USACE in the late 1940s and early 1950s, to move toward a new, more dynamic, method. USACE projects no longer relied solely on hard coastal defense structures as techniques were developed which replicated the protective characteristics of natural beach and dune systems. The resultant use of artificial beaches and stabilized dunes as an engineering approach was an economically viable and more environmentally friendly means for dissipating wave energy and protecting coastal developments. Artificial beaches also had more aesthetic and recreational value then structured shores. The transition from hard structures to beach fill approaches is depicted in Figure I-3-13, which compares the percentage of Federal shore protection funds spent on beach nourishment and on coastal protection structures per decade. Since the 1970s, 90 percent of the Federal appropriation for shore protection has been for beach nourishment (Hillyer 1996).

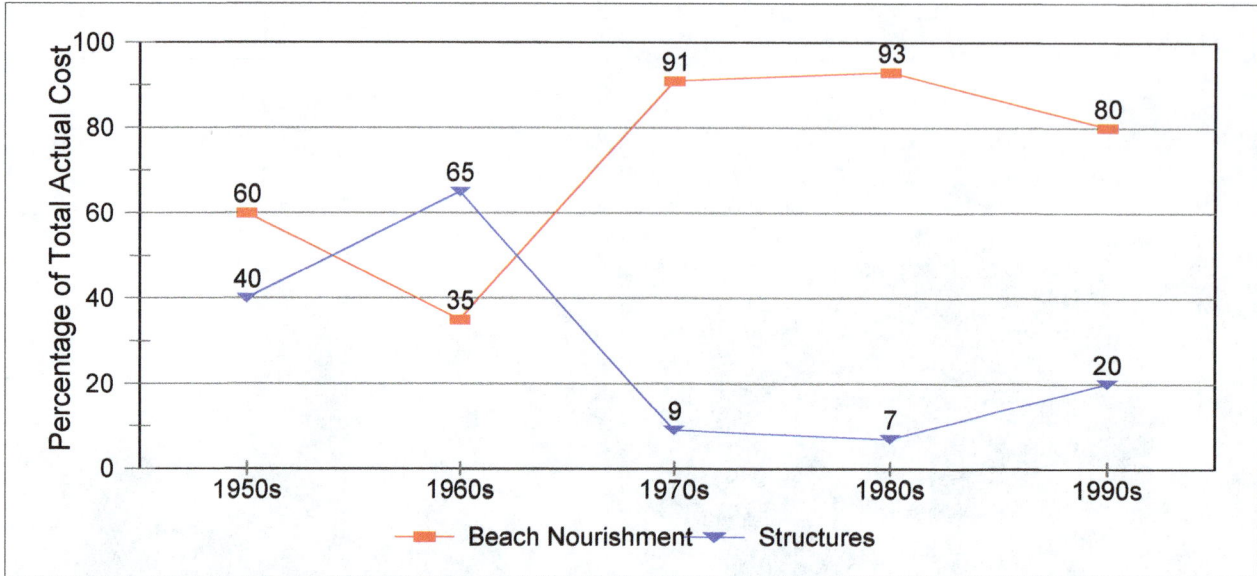

Figure I-3-13. The shift from fixed structures to beach restoration and nourishment (from Hillyer 1996)

l.   The Coastal Engineering Research Center and the Coastal Engineering Research Board. In 1962, the USACE studied the merits of strengthening its coastal engineering research capabilities and the benefits from having the evaluation and reporting of coastal projects follow the same procedures as river, harbor, and flood control investigations. Responding to the recommendations of the Corps' internal study board, Congress, by approving Public Law 172 on November 7,1963, abolished the BEB and established the Coastal Engineering Research Center (CERC). CERC had the same mission as the BEB less its review function, but retained an advisory system in a "Board on Coastal Engineering Research, constituted by the Chief of Engineers in the same manner as the present Beach Erosion Board" (Moore and Moore 1991).

(1) Early years. The Coastal Engineering Research Board (CERB) and CERC followed the lead of their predecessor, the BEB, in pursuing field measurements and basic coastal processes research. The argument was that more research would produce more data, provide for more sound coastal engineering approaches, and lead to greater savings. Spurred by both increasing development and environmental awareness, CERC planned programs to quantify phenomena that previously had been only understood qualitatively. The Marine Science Council, created by the Marine Resources and Engineering Development Act of 1966, appointed the USACE as coordinating agency in a multidisciplinary, interagency effort to identify the effects of construction on the coastal zone. That same year, USACE Headquarters (HQUSACE) asked CERC to draft a program covering the Corps' long range needs in coastal engineering. This triggered a reevaluation and a program increase between 1964 and 1969 (Moore and Moore 1991).

(2) Fort Belvoir. As CERC assumed new missions, its most critical needs were office space and a shelter for the Shore Processes Test Basin (SPTB). Weather conditions limited the open-air SPTB use to the period April through October. A HQUSACE command inspection of CERC in December, 1967, concluded that there was not enough space at the Dalecarlia site for the future CERC. A plan was developed to build a research and development complex in the northwest corner of Fort Belvoir on 182 hectares (450 acres). Several USACE and Department of Army agencies would be located at the complex. CERC, the Board of Engineers for Rivers and Harbors, and the Institute for Water Resources would be located in the Kingman Building. CERC was allocated 40.5 hectares (100 acres) for the replacement of existing test facilities and future expansion.

(3) Field Research Facility. Prior to its move to Fort Belvoir, CERC had planned and budgeted to construct a Field Research Facility (FRF) to evaluate and examine coastal phenomena on prototype (full-size) scale. CERC learned that the U.S. Navy was preparing to surplus a bombing range at Duck, North Carolina, and acquired the property in 1973. On 29 August 1980, the 50th Anniversary of the creation of the Beach Erosion Board, the FRF was officially opened. The 73.7-hectare (182-acre) FRF stretches from ocean to sound, contains a 589-m (1800-foot) pier and laboratory facilities and is used for physical and biological studies (Mason 1979). Meteorological, topographical and oceanographic data are continuously recorded, and the staff conduct research projects at the site and frequently host large field experiments involving other Federal, state and local agencies, plus U.S. and foreign universities. The FRF's advantages of ocean location, research pier, sophisticated infrastructure, synoptic and continuing hydrodynamic and process database, and experienced staff are unique in the United States. Data are accessible on the Internet at the FRF's Web page: http://www.frf.usace.army.mil/frf.html

(4) Shore Protection Manual. When first established, CERC was the only Federal agency with a mission in coastal engineering and almost alone in funding studies of waves and their effects. The research programs at CERC, with their field and laboratory testing and data collection, had an immense practical value. CERC's first combined volume containing guidance on coastal science and engineering was *Shore Protection, Planning, and Design*, Technical Report No. 4 (TR-4), first issued in 1954. The USACE District and Division staffs had a need to apply the data and research results reported by CERC into useful design tools, and they often relied on TR-4 and some related Engineering Manuals published by HQUSACE for design guidance. The *Shore Protection Manual* (SPM) was first published by CERC in 1973 as the updated replacement for TR-4. CERC used the SPM as its primary technology transfer mechanism for many years, with a second edition printed in 1975, a third in 1977, and a fourth and final edition in 1984.

(5) Waterways Experiment Station. A number of events and policy changes in the early 1980's shifted CERC's emphasis into more applied research and moved the laboratory to the Waterways Experiment Station (WES) in Vicksburg, Mississippi. Despite disruptions caused by the 1983 relocation and declining research budgets, CERC prospered in Vicksburg. Reimbursable project work more than compensated for the decline. Mathematical modeling, sophisticated wave tanks and basins (part of the reason for the move), and a closer, more responsive relation with the USACE District and Division staffs all contributed to increased workload.

(6) The Coastal and Hydraulics Laboratory. In the early 1990s, due to political and policy changes, Federal funding for the beach erosion control and flood control projects was severely curtailed and closely regulated. This resulted in reduced research funding and a decrease in the number of new beach erosion control and flood control studies at CERC and the Hydraulics Laboratory. During 1996, both laboratories were combined into one new entity, the Coastal and Hydraulics Laboratory (CHL). CERC's traditional functions such as coastal engineering research, design guidance development such as this manual, and design assistance are still provided by the CHL with the advice of the CERB and a field review group of Division and District staff engineers.

I-3-8 Coastal Engineering in the Military

a. Amphibious operations. Amphibious military operations are not new. Herodotus (1992, translation) describes, in *The Histories*, how Xerxes constructed and used a floating causeway across the Hellespont (the Dardanelles) in 480 B.C. The first amphibious operation in the Americas was the 49-day siege of the French Fortress of Louisburg on Cape Breton Island, Nova Scotia, Canada, in 1745. The Chief Engineer of the operation was Richard Gridley who published that same year the first map in America, a "Plan of the City and Fortifications of Louisburg," and who later became the first Chief Engineer under Commander-in-Chief George Washington in 1775. Many amphibious operations were conducted in North Africa, Italy, France, and the Pacific during World War (W.W.) II. These exercises taught us that for a successful over-the-beach assault, details and forecasts of changes must be acquired of coastal type, beach configuration, morphodynamics, profiles, wave conditions, tides, beach material, beach trafficability, and nearshore and offshore bottom-holding capacity for moorings and anchored ships. At the start of the war, many charts were available showing areas safe for deep-draft navigation and details of land topography, but hardly any of the nearshore areas where assault troops and supplies could be landed. Much of this type of information was collected and evaluated during the war. Of prime importance to military amphibious operations are the wave conditions that can be anticipated. Correlations among wind strength, duration, fetch, and wave height and period were developed in the United States and in the United Kingdom (U.K.) for wave forecast for planning and for operations. The state-of-the-art in military coastal engineering at the end of W.W. II was documented in the *Manual on Amphibious Oceanography*, (University of California at Berkley (UCB), Institute of Engineering Research (IER) 1952) (Wiegel 1999).

b. Expedient harbors. Expedient harbor design for the invasion of Normandy also required substantial coastal engineering effort. The design of the two Mulberry harbors ("A" at St. Laurent (Omaha Beach) and "B" at Arromanches) required information on wave and tide prediction (design tide range was 7.3 meters (24 feet)), wave diffraction, wave induced forces, bottom conditions, and placement of structures and their foundations. Wave-diffraction theory (wave transmission about the tip or through a gap between breakwaters) was developed for this project. The Mulberry Harbor was designed in two parts. The portion closest to shore, in shallow water, had a breakwater of vertical reinforced-concrete caissons (code name "Phoenix") and sunken ships protecting it, while the seaward portion was protected by moored floating breakwaters ("Bombardon"). The Bombardon had a cross section similar to a Maltese cross in shape; each unit was 61 meters (200 feet) in length, 7.6 meters (25 feet) in beam and depth with 5.8 meters (19 feet) draft. They were deployed in pairs with a 15.2 meters gap between pairs. Located inside the shallow water sheltered area were pier heads and mile-long pontoon-supported flexible bridges (causeways code named "whales"). After initial construction, a storm along the Normandy coast with gale winds blowing from the northeast generated sea conditions larger than project design waves. Operations were disrupted and delayed, with great damage to the artificial harbors, craft and ships. Mulberry "A" suffered damage beyond repair. Shown in Figure I-3-14 Is Mulberry "B" after being repaired. *The Civil Engineer in War, A Symposium of Papers on Wartime Engineering Problems, Volume 2, Docks and Harbors* (Institute of Civil Engineers 1948) provides details on the design, installation, and performance of the Mulberry Harbors (Wiegel 1999).

c. Military coastal engineering studies. After W.W. II, UCB contracted with the Office of Naval Research (ONR) to review amphibious operations reports from the war. As expected, many landing-craft and amphibious-vehicle casualties were due to enemy action, but many were related to problems with waves and currents causing capsizing, swamping, broaching, getting stuck on bars and, when the ramps were down, filling with water and sand. Another major problem was beach trafficability. Vehicles were frequently stuck in the sand. A trafficability study of beach sand characteristics, beach slope, water level, and vehicle type was made. It was observed that saturated sand near the water's edge would liquefy due to vibrations produced by the vehicular traffic. Several full-scale amphibious assault-training exercises were observed in detail and reports prepared on the observations and findings.

During the 25 October, 1949, exercise across three west coast beaches at the Waianae-Pokai Bay region of Oahu, Hawaii, long-period waves surging up the steep beach face caused substantial landing craft casualties on two of the beaches. Many of the craft broached and were shoved onto the steep beach by the surging breakers (see Figure I-3-15). Of the 20 landing craft sent ashore in 3 "waves" in the first 15 minutes of the amphibious exercise, 7 retracted and 8 were lost, some filling with water and sand when the ramps were lowered. The exercise was quickly halted and five of the craft later salvaged. Because of the problems experienced moving personnel, equipment and supplies through the surf and over the beach, the Department of Defense began the development of helicopters and air cushion vehicles (Wiegel 1999).

Figure I-3-14. Aerial photograph of the Mulberry "B" Harbor at Arromanches, France, June 1944 (from Institution of Civil Engineers, London, England, 1948, p. 242)

d. Port operations, Republic of Korea. One of the major amphibious operations in the Republic of Korea (ROK) was the invasion at Inchon Harbor. The east coast of the ROK is generally rocky sea cliffs, while the south and west coasts are extensive mud flats with many small conical islands. During low tide at Inchon, the tidal flats extend at least 50 km offshore and the only approach to the port is a narrow channel 1 to 2 meters deep at zero tide. Extreme tides, ranging from -0.6 meters to +9.8 meters, provided many engineers their first experience with a sea of water quickly changing into a sea of mud. After the 15 September 1950 amphibious assault, General MacArthur said "... conception of the Inchon landing would have been impossible without the assurance of success afforded by the use of the Seabee pontoon causeways and piers." The tidal basin and other facilities were relatively intact at the time of the invasion. However, during the U.S. evacuation in January 1951, the lock gates to the tidal basin were demolished, then rebuilt after return of the U.S. forces in February 1951 (Wiegel 1999).

e. Port operations, Republic of Vietnam. In Vietnam, the U.S. Navy Officer in Charge of Construction was responsible for Saigon-Newport and DaNang deep-water ports. Data were collected on coastal sediments, storms, tides, and waves to plan the dredging of these ports and seven other sites. The Army port construction section used these data and the Beach Erosion Board's Technical Report No. 4, *Shore Protection Planning and Design*, to prepare detailed design of port facilities. The DeLong Corporation's installation of piers, prefabricated causeway components, and use of self elevating barges also contributed to successful port facility operations. The sand on many of the Vietnamese beaches was of such character that over-the-shore operations were almost impossible. Rock, concrete, articulators, pierced-steel planking and other means attempted to stabilize the sand were serviceable for very limited periods ranging from days to two weeks. Waves on the foreshore undermined the stabilizing structures and decreased the bearing capacity, resulting in the structures sinking out of sight. Engineers, drawing on their W.W. II and subsequent experience, used draglines and blasting to dig nearby coral, crush it, and then place it in compacted layers where the foreshore had been previously excavated for the purpose. This installation lasted several months with only minor repair and was judged very satisfactory (Wiegel 1999).

Figure I-3-15. Operation MIKI, Waianae, Oahu, HI, October 1949, (Many landing craft broached and were shoved onto the beach by surging breakers) (from Wiegel and Kimberly 1949)

(1) Vung Ro. For the construction of the port of Vung Ro to support the major air base at Tuy Hoa, estimates of typhoon and monsoon waves were made and wave refraction studies used to plan the port. Port development included landing-craft ramps, a floating pontoon cube barge dock, two DeLong barge piers (183 x 24.4-meters) (600 x 80-feet) with two girder spans off a 120-meter (394-foot) rock fill causeway, tanker mooring facilities and a pipeline connecting to the tank farm south of Tuy Hoa airstrip about 15.5 km (9.6 miles) distance. Seven months after the port was completed the facilities were struck by a typhoon, but survived wave conditions exceeding the design wave criteria (Wiegel 1999).

(2) Da Rang River. The U.S. Air Force wanted a harbor closer to the air base that would be located at the mouth of the Da Rang River. The river brings a great amount of sand to the coast in this area. The USACE had recommended, based on studies by CERC and WES, a site several miles south where the 3-fathom line was much closer to shore. The Navy had funded hydraulic model studies of the entrance and jetties at the southern site. However, the Air Force pressed on at the river mouth site with a "turnkey project" contractor. After several months of pilot channel dredging, the contractor conceded that construction of a harbor at the Da Rang River mouth was impractical and the project was abandoned (Wiegel 1999).

f. Temporary wharfs. The Army has two sizes of self-elevating pier systems, known as DeLong barges. These units are honeycomb-like, welded-steel, box-girder structures consisting of plates and stiffeners that are towed to a site where a temporary harbor is being built or an existing port rehabilitated. The units are anchored on pneumatic caissons that are forced into the sea bed with air jacks. The units can

History of Coastal Engineering

be used on noncohesive beds but are unsuitable for rock, organic materials, or wet clays and silts. The units can be installed in a variety of configurations at the site (Figure I-3-16) (Department of the Army 1990).

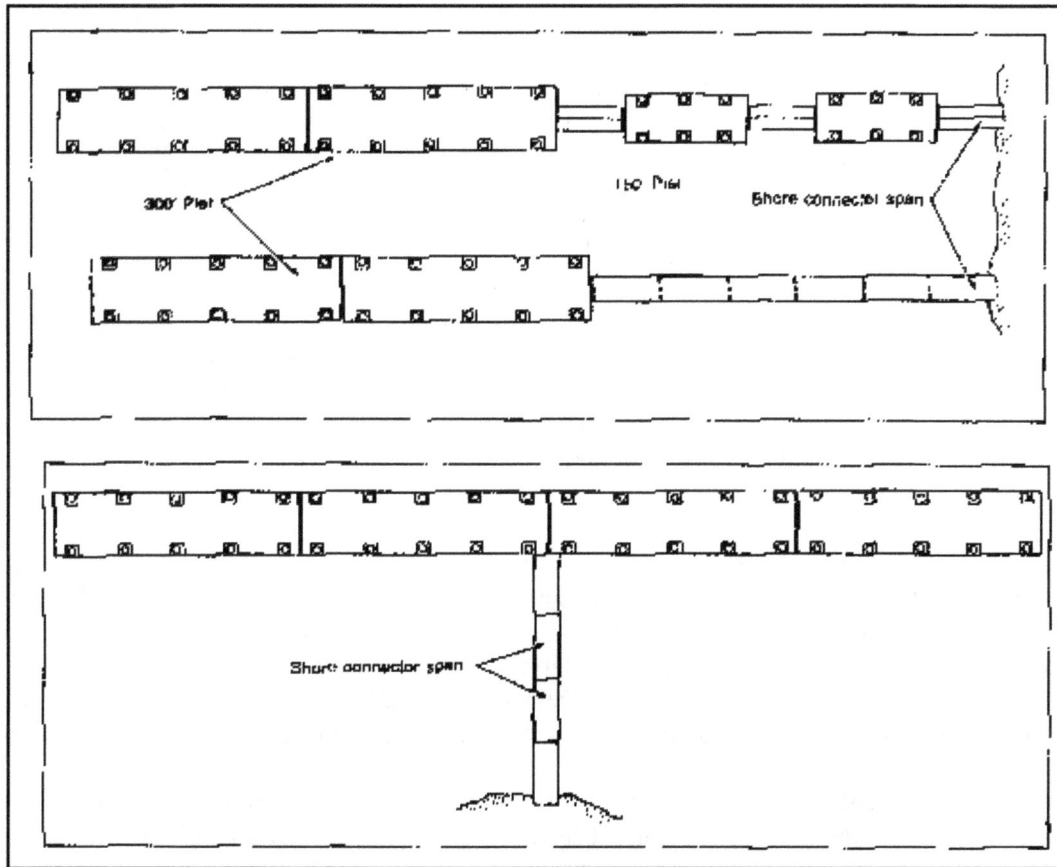

Figure I-3-16. DeLong self-elevating barge piers. Upper figure shows finger piers, lower a T-type marginal wharf (from Department of the Army 1990)

g. Rapidly Installed Breakwater System. Engineers and scientists at CHL are developing a Rapidly Installed Breakwater (RIB) system to address problems encountered by the U.S. armed forces while offloading ships during Logistics Over The Shore operations. When seas become sufficiently energetic during offloading, the capabilities of ship-based crane operators and stevedore crews are severely restricted. The RIB system, consisting of a series of floating breakwater units that are assembled in a "V" orientation, is designed to create a "pool" of calmer water where the crews will be able to continue to unload vessels even during storms. For many years, CHL had been involved with the design and deployment of floating breakwaters, primarily for application within bays or estuaries which are semi-protected from large waves. But, these structures were intended to attenuate waves with heights not exceeding 4 ft and periods not exceeding 4 sec, while in an oceanic environment, waves with heights up to 10 ft are common during storms, with associated periods up to 10 sec. To date, research efforts have concentrated on military applications for the RIB system. Potential civil applications include rescue and recovery operations, temporary small vessel shelter from energetic seas, and to protect exposed dredging and marine operations (e.g., bridge repair, rubble-mound breakwater construction),.

I-3-9. Summary

For most of the nation's history, the U.S. Army Corps of Engineers has played an active role in the coastal zone. To the mid-1800s, this role was largely confined to coastal defense and some harbor protection. But, in the mid-1800s, the USACE's mission expanded to include developing civil works projects in support of a growing nation. These responsiblities included harbor construction, dredging and clearing waterways, building canals and channels, and protecting coastal areas threatened by erosion (*e.g.*, Presque Isle). During the second half of the 20th century, the USACE's role further expanded to include environmental restoration and preservation of threatened coastal areas. Since the 1930s, coastal-related research and development have been conducted to advance the technical foundations and basis for conducting coastal civil works.

The 20th century was witness to a large-scale evolution in the development of, use of, and interest in the coastal zone. National defense, agriculture, navigation, economic development, recreation, and environmental worth all contribute to the definition of coastal policy and action. During the early years of the 21st century there will be continuing development pressure in the coastal zone .   Coastal engineers and scientists will undoubtedly be asked to play an increasing role in planning, designing, and maintaining infrastructure projects, in coastal management and environmental mitigation, and will continue their more traditional missions of navigation and flood protection.

In fiscal year 1998, the USACE and contractor-owned dredges removed 182 million cubic meters of material from Federally-constructed and maintained channels at a cost of $713 million. Dredged material is a valuable resource with numerous potential benefits, including construction of protective dunes and beaches, maintenance of beaches through bypassing to reestablish natural sediment-transport paths, and restoration and creation of wetlands and coastal habitat. Demand for dredged material usage is increasing, but environmental concerns and constraints present new engineering challenges that must be addressed.

Erosion and flooding threaten an estimated $3 trillion of development along the coast, with 80 to 90 percent of the nation's sandy beaches eroding (Hillyer 1996). Shore protection and restoration throughout the developed areas of the coast will increase, especially if the growing value of coastal property and recreation benefits are factored into the cost benefit calculations.

Because of the age of many harbor structures, improving and rebuilding jetties and breakwaters will be a major mission area. Wetlands restoration should also be growth areas, and already the USACE is involved in major restoration projects in the Everglades, in south Louisiana, and along many stretches of the intracoastal waterway.

Emergency coastal response work is also likely to be a growth area for the USACE. Many of the recent arrivals to the coastal zone have not personally experienced a major disaster like the Galveston hurricane of 1900, the 1962 Ash Wednesday storm, or the Great New England Hurricane of 1938. Much of the population is ignorant of the hazards that exist and is not prepared to respond to the aftermath of a catastrophic storm. The USACE has actively participated in disaster emergency and recovery efforts in Puerto Rico and other Caribbean islands, and many of these skills are applicable to mainland disasters.

I-3-10. References

Alperin 1977
Alperin, L. M. 1977. *Custodians of the Coast: History of the United States Army Engineers at Galveston.* Galveston District, U.S. Army Corps of Engineers, Galveston, TX, 318 p.

Barber 1989
Barber, H. E. 1989. *A History of the Savannah District, U.S. Army Corps of Engineers.* Savannah District, U.S. Army Corps of Engineers, Savannah, GA, 554 p.

Becker 1984
Becker, W. H. 1984. *From the Atlantic to the Great Lakes, a history of the U.S. Army Corps of Engineers and the St. Lawrence Seaway.* Office of the Chief of Engineers, Washington, DC. 179 p.

Bijker 1996
Bijker, E. W. 1996. History and Heritage in Coastal Engineering in the Netherlands. *History and Heritage of Coastal Engineering,* N. C. Kraus, ed., Coastal Engineering Research Council, American Society of Civil Engineers, Reston, VA, pp. 390 - 412.

Birkemeier, Dolan, Fisher 1984
Birkemeier, W. A., Dolan, R., and Fisher, N. 1984. The Evolution of a Barrier Island: 1930-1980. *Shore and Beach*, Vol. 52, No. 2, pp 2-12.

Buker 1981
Buker G. E. 1981. *Sun, Sand and Water: a History of the Jacksonville District U.S. Army Corps of Engineers, 1821-1975.* U.S. Government Printing Office, Washington, DC., 288 p.

Chambers 1980
Chambers, J. W., II. 1980. *The North Atlantic Engineers : a History of the North Atlantic Division and its Predecessors in the U.S. Army Corps of Engineers, 1775-1975.* North Atlantic Division, U.S. Army Corps of Engineers, New York, NY, 167 p.

Chrzastowski 1999
Chrzastowski, M. J. 1999. Geology of the Chicago Lakeshore, Shaping the Chicago Shoreline, poster produced by the Department of Natural Resources, Illinois State Geological Survey, Champaign, IL.

Cowdrey 1977
Cowdrey, A. E. 1977. *Land's End: a History of the New Orleans District, U. S. Army Corps of Engineers, and its Lifelong Battle with the Lower Mississippi and Other Rivers Wending Their Way to the Sea.* New Orleans District, U.S. Army Corps of Engineers, New Orleans, LA, 118 p.

Davis 1975
Davis, V. S. 1975. *A History of the Mobile District 1815 to 1971.* Mobile District, U.S. Army Corps of Engineers, Mobile, AL, 109 p.

Department of the Army 1990
Department of the Army. 1990. Port Construction and Repair. Field Manual FM 5-480, Headquarters, Department of the Army, Washington, DC.

DeWan 1999
DeWan, G. 1999. "The Master Builder, How planner Robert Moses transformed Long Island for the 20th Century and Beyond." *Long Island, Our Story.* http://www.lihistory.com/ (18 May 2000).

Dobney 1978
Dobney, F. J. 1978. *River Engineers on the Middle Mississippi: a History of the St. Louis District, U.S. Army Corps of Engineers.* U.S. Government Printing Office, Washington, DC., 177 p.

Drescher 1982
Drescher, N. M. 1982. *Engineers for the Public Good, A History of the Buffalo District, U.S. Army Corps of Engineers,* U. S. Army Corps of Engineers, Buffalo, NY, pp 155-173, 203-235.

Empereur 1997
Empereur, J. Y. 1997. "The Riches of Alexandria." Transcript of a 1997 Interview on NOVA Online, http://www.pbs.org/wgbh/nova/sunken/empereur.html (28 Aug 2000).

Farley 1923
Farley, P. P. 1923. Coney Island public beach and boardwalk improvement. *The Municipal Engineers Journal*, Vol. 9, Paper 136, pp 136.1-136.32.

Fasso 1987
Fasso, C. A. 1987. "Birth of hydraulics during the Renaissance period," Hydraulics and Hydraulic Research; a Historical Review, IAHR, G. Garbrecht Editor, Balkema, pp 55-79

Franco 1996
Franco, Leopold. 1996. "History of Coastal Engineering in Italy," *History and Heritage of Coastal Engineering,* American Society of Civil Engineers, New York, NY. pp 275-335.

Frost 1963
Frost, H. 1963. *Under the Mediterranean, Marine Antiquities.* Prentice-Hall, Englewood Cliffs, N.J. p 278.

Goreki and Pope 1993
Goreki, R. J., and Pope, J. 1993. Coastal Geologic and Engineering History of Presque Isle Peninsula, Pennyslvania. Miscellaneous Paper CERC-93-8, U.S. Army Engineer Waterways Experiment Station, Vicksburg, MS.

Hagwood 1982
Hagwood, J. J., Jr. 1982. *Engineers at the Golden Gate: a History of the San Francisco District, U.S. Army Corps of Engineers, 1866-1980.* San Francisco District, U.S. Army Corps of Engineers, San Francisco, CA, 453 p.

Hartzer 1984
Hartzer, R. B. 1984. *To Great and Useful Purpose: a History of the Wilmington District U.S. Army Corps of Engineers,* Wilmington District, U.S. Army Corps of Engineers, Wilmington, NC, 172 p.

Herodotus 1992
Herodotus. 1992. *The Histories,* Written in the middle of the 5th century B.C., translation by George Rawlinson and introduction by Hugh Bowden, J. M. Dent & Sons Ltd.,London.

Hillyer 1996
Hillyer, T. M. 1996. *Shoreline Protection and Beach Erosion Control Study, Final Report: An Analysis of the U.S. Army Corps of Engineers Shore Protection Program.* IWR Report 96-PS-1, Shoreline Protection and Beach Erosion Control Task Force, Water Resources Support Center, Institute for Water Resources, U.S. Army Corps of Engineers, Alexandria, VA (prepared for the Office of Management and Budget, Washington, DC.)

Inman 1974
Inman, D. L. 1974. Ancient and Modern Harbors: a Repeating Phylogeny. *Proceedings of the 14th Coastal Engineering Conference,* Copenhagen, Denmark, June 1974, American Society of Civil Engineers, Reston, VA, pp. 2049-2067.

Inman 2001
Inman, D. L. 2001. History of Early Breakwaters. *Association of Coastal Engineers Newsletter*, Alexandria, VA.

Institute of Civil Engineers 1948
Institution of Civil Engineers. 1948. *The Civil Engineer in War. A Symposium of Papers on War-time Engineering Problems. Vol. 2, Docks and Harbors.* Papers and discussions, published by The Institution of Civil Engineers, London, United Kingdom.

Jacobs 1976
Jacobs W. A. 1976. *The Alaska District, United States Army, Corps of Engineers, 1946-1974: a History.* From a manuscript researched and compiled by Lyman L. Woodman, Alaska District, U.S. Army Corps of Engineers, Elmendorf AFB, Alaska, 145 p.

Johnson 1979
Johnson, L. R. 1979. *The Headwaters District: a History of the Pittsburgh District, U.S. Army Corps of Engineers.* Pittsburgh District, U.S. Army Corps of Engineers, Pittsburgh, PA, 380 p.

Johnson 1992
Johnson, L. R. 1992. *The Ohio River Division, U.S. Army Corps of Engineers, the History of a Central Command.* Ohio River Division, U.S. Army Corps of Engineers, Cincinnati, OH, 484 p.

Kana 1999
Kana. T. 1999. Long Island's South Shore Beaches, A Century of Dynamic Sediment Management. *Coastal Sediments '99,* Kraus, N. C., and McDougal, W. G. (eds.), American Society of Civil Engineers, Reston, VA, pp.1584-1596.

Karanek 1978
Kanarek, H. K. 1978. *The mid-Atlantic engineers: a History of the Baltimore District, U.S. Army Corps of Engineers, 1774-1974.* U.S. Government Printing Office, Washington, DC., 196 p.

Keay 1942
Keay, T. B. 1942. Coast Erosion in Great Britain, General Question of Erosion and Prevention of Damage; and the Drainage of Low-Lying Lands. *Shore and Beach,* Vol. 10, No. 2, pp. 66 - 68.

Klawonn 1977
Klawonn, Marion J. 1977. *Cradle of the Corps, A History of the New York District, U.S. Army Corps of Engineers, 1975-1975,* New York District, U.S. Army Corps of Engineers, New York, NY, 310 p.

Lamb 1930
Lamb, H. 1930. *The Crusades, Iron Men and Saints,* Doubleday, Doran & Co., Garden City, NY.

Larson 1979
Larson, J. W. 1979. *Those Army Engineers: a History of the Chicago District, U.S. Army Corps of Engineers,* Chicago District, U.S. Army Corps of Engineers, Chicago, IL, 307 p.

Larson 1981
Larson, J. W. 1981. *Essayons: a History of the Detroit District, U.S. Army Corps of Engineers.* Detroit District, U.S. Army Corps of Engineers, Detroit, MI, 215 p.

Lenček and Bosker 1998
Lenček, L., and Bosker, G. 1998. *The Beach: The History of Paradise on Earth.* Viking, New York, 310 p.

Maass 1951
Maass, A. 1951. *Muddy Waters, the Army Engineers and the Nation's Rivers.* Harvard University Press, Cambridge, MA, 306 p.

Mason 1979
Mason, Curtis. 1979. "The Coastal Engineering Research Center's Field Research Facility at Duck, North Carolina," *Shore & Beach, Journal of the American Shore and Beach Preservation Association,* Vol. 47, No. 2, pp 13-16.

Merritt 1979
Merritt, R. H. 1979. *Creativity, Conflict & Controversy: a History of the St. Paul District, U.S. Army Corps of Engineers.* U.S. Government Printing Office, Washington, DC., 461 p.

Moore 1981
Moore, 1981. *The Lowcountry Engineers: Military Missions and Economic Development in the Charleston District,* U.S. Army Corps of Engineers. U.S. Government Printing Office, Washington, DC., 140 p.

Moore and Moore 1982
Moore, Jamie W., and Moore, Dorthy P. 1982. *The Lowcountry Engineers, Military Missions and Economic Development in the Charleston District,* U.S. Army Corps of Engineers (Atlanta, GA: Government Printing Office), pp16-21

Moore and Moore 1983
Moore, Jamie W., and Moore, Dorthy P. 1983. The Corps of Engineers and Coastal Engineering, A 50 Year Retrospective,? Orville T. Magoon, ed., *Coastal Zone 83 III,* New York: ASCE, 1983, pp 1627-1639

Moore and Moore 1991
Moore, Jamie W. and Moore, Dorthy P. 1991. *History of the Coastal Engineering Research Center, 1963-1983,* U. S. Army Engineer Waterways Experiment Station, Vicksburg, MS

Morison and Commager 1962
Morison, S. E., and Commager, H. S. 1962. *The Growth of the American Republic.* Fifth Ed., Oxford University Press, New York.

Palmer and Tritton., eds. 1996
Palmer, R., and Tritton Limited (eds.). 1996. History of Coastal Engineering in Great Britain. *History and Heritage of Coastal Engineering,* N. C. Kraus, ed., Coastal Engineering Research Council, American Society of Civil Engineers, Reston, VA, pp. 214 - 274.

Parkman 1978
Parkman, Aubrey. 1978. *Army Engineers in New England, The Military and Civil Works of the Corps of Engineers, 1775-1975,* Waltham, MA, U.S. Army Engineer Division, 1978, pp34-38.

Parker 1980
Parker 1980. "Structures in the Beach Zone-To Be or Not to Be," paper presented at the American Soc. Of Civil Engineers Convention and Exposition, Florida, 27-1 Oct, 1980, ASCE Reprint No.80-648.

Patton 1934
Patton, R. S. 1934. "The Purpose of the American Shore and Beach Preservation Association," *Shore and Beach II,* Journal of the American Shore and Beach Preservation Association Oct. 1934, pp132-136.

Quinn 1972
Quinn, Alonzo DeF. 1972. *Design and Construction of Ports and Marine Structures,* Second Edition, Mc Graw-Hill Book Company, New York, N.Y., 611p.

Rathburn 1990
Rathbun, M. Y. 1990. *Castle on the Rock, 1881-1985: the History of the Little Rock District, U.S. Army Corps of Engineers.* Little Rock District, U.S. Army Corps of Engineers, Little Rock, AK, 192 p.

Reuss 1998
Reuss, M. 1998. *Designing the Bayous: the Control of Water in the Atchafalaya Basin: 1800-1995.* Office of History, U.S. Army Corps of Engineers, Alexandria, VA, 474 p.

Richter 1970
Richter, J. P. 1970. *The Notebooks of Leonardo da Vinci,* Dover Publishing, New York, NY.

Rouse and Ince 1963
Rouse, H. and Ince, S. 1963. *History of Hydraulics,* Dover, NY.

Seattle District 1969
Seattle District. 1969. *History of the Seattle District, 1896-1968 : United States Army, Corps of Engineers : 72 Years in Peace and War.* Seattle District, U.S. Army Corps of Engineers, Seattle, WA, 222 p.

Shore Protection Manual 1984
Shore Protection Manual. 1984. 4th ed., 2 Vol., U.S. Army Engineer Waterways Experiment Station, U.S. Government Printing Office, Washington, DC.

Shore Protection, Planning, and Design 1954
Shore Protection, Planning, and Design. 1954. Technical Report No. 4, U.S. Beach Erosion Board, Washington, DC, 393 p.

Snyder and Guss 1974
Snyder, F. E. and Guss, B. H. 1974. The District; a History of the Philadelphia District, U.S. Army Corps of Engineers, 1866-1971, U.S. Army Engineer District, Philadelphia, PA, 263 p.

Stanton 1999
Stanton, J. 1999. "Coney Island - Nickel Empire (1920's-1930's)." *Coney Island History Site.* North American Integration & Development Center, School of Public Policy and Social Research, University of California, Los Angeles, CA, http://naid.sppsr.ucla.edu/coneyisland/ (16 May 2000).

Straub 1964
Straub, Hans. 1964. *A History of Civil Engineering,* English Translation by Erwin Rockwell, The M.I.T. Press, Massachusetts Institute of Technology, Cambridge, Massachusetts, 258 p.

Sudar, Pope, Hillyer, and Crumm 1995
Sudar, R. A., Pope, J., Hillyer, T., and Crumm, J. 1995. "Shore Protection Projects of the U.S. Army Corps of Engineers." *Shore and Beach.* Journal of the American Shore and Beach Preservation Association, Vol. 63, No. 2, pp. 3-16.

Thompson 1981
Thompson, E. N. 1981. *Pacific Ocean Engineers-History of the U.S. Army Corps of Engineers in the Pacific, 1905-1980.* U.S. Army Engineer Division, Pacific Ocean, Honolulu pp 460.

Turhollow 1975
Turhollow, Anthony F. 1975. *A History of the Los Angeles District, U. S. Army Corps of Engineers, 1898-1965,* Los Angeles, CA, U. S. Army Engineer District, pp 20-48.

UCB, IER 1952
University of California, Berkeley, Institute of Engineering Research. 1952. *Manual on Amphibious Oceanography,* R. L. Wiegel, Project Engineer and Editor, under contract with Office of Naval Research, Amphibious Branch, Contract N7-29535, Project No. NR 252-003, Pentagon Press, July 1952, 2 volumes.

USACE 1971
U.S. Army Corps of Engineers. 1971. National Shoreline Study: Shore Protection Guidelines. Department of the Army, Corps of Engineers, Washington, DC.

USACE 1980
U.S. Army Corps of Engineers. Army 1980. *Annual Report of Civil Works Activities of the Chief of Engineers for FY 1979, Volume 1, Summary/Highlights,* Department of the, Office of the Chief of Engineers, Washington, DC. 12 August 80, pp 44-47.

U.S. Army Corps of Engineers 1998
U.S. Army Corps of Engineers. 1998. *The History of the US Army Corps of Engineers.* Engineer Pamphlet EP 360-1-21, Washington, DC.

Waterborne Commerce Statistics Center 1999
Waterborne Commerce Statistics Center, Water Resources Support Center, Navigation Data Center, U.S. Army Corps of Engineers, Alexandria, VA. http://www.wrsc.usace.army.mil/ndc/ (6 Dec 1999).

Wilson and Eaton 1960
Wilson, MG Walter K., Jr. and Eaton, Richard O. 1960. "A History of the Beach Erosion Board," *Shore and Beach,* Journal of the American Shore and Beach Preservation Association, Volume Twenty-eight, Number one, April 1960, pp 4-9.

Wiegel 1999
Wiegel, Robert L. 1999. *Military Examples of Coastal Engineering,* Miscellaneous Paper CHL-99-3, September 1999, Prepared for Coastal Engineering Research Board, U.S. Army Corps of Engineers Research and Development Center, Vicksburg, MS 39180-6199.

Wiegel and Kimberly 1949
Wiegel, Robert L., and Kimberly, H. L. 1949. *Operation MIKI, Hawaiian amphibious phase (ground coverage),* University of California, Berkley, Inst. Engrg. Res., Series 29, Issue 14.

Wiegel and Saville 1996
Wiegel, R. L., and Saville, T., Jr. 1996. History of Coastal Engineering in the United States. *History and Heritage of Coastal Engineering,* N. C. Kraus, ed., Coastal Engineering Research Council, American Society of Civil Engineers, Reston, VA, pp. 214 - 274.

Willingham 1983
Willingham, W. F. 1983. *Army Engineers and the Development of Oregon, a History of the Portland District U.S. Army Corps of Engineers.* Portland District, U.S. Army Corps of Engineers, Portland, OR.

van Hoften 1970
van Hoften, E. 1970. *History of the Honolulu Engineer District, 1905-1965.* U.S. Army Engineer District, Honolulu, pp 133.

Vitruvius, Rowland, and Howe, eds. 1999
Vitruvius, P., Rowland, I. D., and Howe, T. N. eds. 1999. Vitruvius: Ten Books on Architecture. Cambridge University Press, Cambridge, MA.

I-3-11. Acknowledgments

Authors of Chapter I-3, "History of Coastal Engineering:"

John H. Lockhart. Jr., Headquarters, U.S. Army Corps of Engineers, Washington, DC, (retired).
Andrew Morang, Ph.D., Coastal and Hydraulics Laboratory (CHL), Engineer Research and Development
    Center, Vicksburg, Mississippi.

Reviewers:

Joan Pope, CHL
Martin A. Reuss, Ph.D., Office of History, U.S. Army Corps of Engineers, Alexandria, Virginia.

Table of Contents

Chapter I-4
The Coastal Engineering Manual

I-4-1. Background

During the 1970s, '80s, and '90s, coastal engineering practice by the U.S. Army Corps of Engineers (USACE) and standard engineering for most coastal projects throughout the world have been based, wholly or in part, on the *Shore Protection Manual (SPM)*. Since the *SPM* was last updated in 1984, the coastal engineering field has witnessed many technical advances and increased emphasis on computer modeling, environmental restoration, and project maintenance applications. The BEB produced the first standardized guidance on coastal structure design in 1954, *Shore Protection Planning and Design*, also known as *TR-4*. This was the forerunner of the *SPM* that was first published by CERC in 1973, and revised in 1975, 1977, and 1984. These documents present the methodology that guided coastal structure and beach fill design for most of the projects constructed to date. The USACE traditionally is responsible for constructing and maintaining United States Federally authorized coastal civil works projects including harbor entrance channels, navigation channels and structures, coastal storm damage reduction and shore protection projects. Therefore, the USACE is primarily responsible for developing the principles of coastal engineering as they are practiced in the United States.

a. Shore Protection Planning and Design, TR 4. The methodologies of *TR-4* emphasized designing coastal structures for stability against wave forces. The technology available at that time provided little means to address the functional performance of structures, nor provide any guidance for predicting the performance or stability of a beach fill. Beach and dune design was only qualitatively addressed. Simple linear wave theory, static terrestrial structural engineering principles, and trail-and-error experiential data were used to develop the empirical relationships and rules-of-thumb presented in *TR-4*. Beach fills of this era were not usually designed to perform a particular function, but were typically placed as an added feature to increase the sediment supply in the area of interest and to reduce wave energy striking the protective structures (the primary project feature).

b. Shore Protection Manual, SPM. The *SPM* was a significant advancement over *TR-4* in that it used the results of physical model tests to develop principles of wave-structure interaction, advancements in wave theory, and statistics and other data from various projects. The *SPM* provided significantly more guidance in the positioning and intent of groins and breakwaters, predicting the flood control benefits of seawalls, and predicting the stability of beach fills. At 1,160 pages, the first edition of the *SPM* was almost three times the length of the 20-year-older *TR-4* (Camfield 1988). The *SPM* and beach fill projects of the 1970s and early '80s were designed around the objective of beach erosion control and recreational use. The quantity of material to be placed was computed based on the long-term recession rates, and the amount of surface area desired to support recreational needs. The *SPM* presented guidance to assist in predicting maintenance nourishment quantities based on the grain size of the placed fill and its projected stability relative to the native material grain size. Neither the *SPM* nor the projects constructed during this time concerned themselves with the performance of the beach fill template during a particular storm. At that time, beach fills were not usually designed with a primary purpose of providing flood control benefits.

The *SPM* is commonly used as a university textbook and as a training aid for apprentice engineers. It is also a convenient reference for empirical procedures to compute a particular design parameter. Approximately 30,000 copies have been sold through the U.S. Government Printing Office. Translations into other languages, including Chinese and Catalonnian (Spanish), further attest to the *SPM's* role as an international standard guidance for professional coastal engineers (Pope 1993, 1998). Even though the *SPM* is a general

selecting and using various planning and design tools as appropriate for the project at hand. The engineering tools are presented in a modular grouping to allow for future updates as the technology continues to advance.

d. Part V. "Coastal Project Planning and Design" starts with chapters discussing the planning and design process and site characterization. Following these general chapters are ones discussing the planning and design of shore protection projects (including coastal armoring, beach restoration, beach stabilization and coastal flood protection projects), beach fill, navigation projects (including defining the fleet, entrance channel, inner harbor elements, structures, sedimentation, maintenance, and management), and environmental enhancement projects (including laws, regulations, and authorities, issues, alternative approaches, planning, and design). A final chapter outlines conditions and regulations unique to USACE projects in the United States.

e. Part VI. "Design of Coastal Project Elements" includes chapters discussing philosophy of coastal structure design, the various types and function of coastal structures, site conditions, materials, design fundamentals, reliability, and the design of specific project elements (including a sloping-front structure, vertical-front structure, beach fill, floating structure, pile structure, and a pipeline and outfall structure.

f. Appendix A. The "Glossary of Coastal Terminology" has been compiled from numerous sources and lists terms found throughout the CEM. Note that there is no single, comprehensive list of mathematical terms and symbols. Each CEM chapter has its own symbol list.

g. Updates. The CEM is intended to be a "living document" and to be updated periodically as advances in the field render the existing chapters obsolete or inadequate. Comments and suggestions should be addressed to the Coastal and Hydraulics Laboratory, CEERD-HN-CE. Corrected or modified chapters will be posted on the CHL web page.

I-4-3. References

Camfield 1988
Camfield, F. E. 1988. "Technology Transfer – The Shore Protection Manual," *Journal of Coastal Research,* 4(3), pp 335-338.

Pope 1993
Pope, J. 1993. "Replacing the SPM: The Coastal Engineering Manual." *The State of the Art of Beach Nourishment, Proceedings, 6th Annual National Conference on Beach Preservation Technology, Florida Shore and Beach Preservation Association, Tallahassee, FL,* pp 319-334.

Pope 1998
Pope, J. 1998. "Replacing the SPM: The Coastal Engineering Manual." *PIANC Bulletin,* No. 97, pp 43-46.

USACE 1954
USACE 1954. *Shore Protection Planning and Design, Technical Report No. 4,* Beach Erosion Board, U.S. Government Printing Office, Washington, DC.

***Shore Protection Manual* 1984**
Shore Protection Manual, 4th ed., 2 Vol., U.S. Army Engineer Waterways Experiment Station, U.S. Government Printing Office, Washington, DC, 1,088 p.

coastal engineering reference, some aspects of navigation and harbor design are not included and its primary focus is shore protection.

c.   Coastal Engineering Manual, CEM. The advent of numerical models, reliable field instrumentation techniques, and improved understandings of the physical relationships which influence coastal processes lead to more sophisticated approaches in shore protection design in the later 1980s and 90s. Numerous guidance and analytical tools have been developed over the last 15 years to assist the coastal engineer in predicting not only the stability of a beach fill, but also its performance during extreme events. Cross-shore and alongshore change models, hydrodynamic hind cast data bases, and stochastic statistical approaches have been developed to provide the practicing coastal engineer with procedures for quantifying the flood control benefits of a proposed design. The functional interaction of beach erosion control structures (i.e., groins and breakwaters) can be analyzed with numerical simulation. Seawalls can be designed not only for stability, but also physically modeled to predict various elements of the wave-structure interaction including scour and overtopping. A "modern" technical document incorporating all the tools and procedures used to plan, design, construct, and maintain coastal projects was needed. The USACE tasked the Coastal Engineering Research Center and, later, the Coastal and Hydraulics Laboratory with producing a new reference incorporating established science and much of this new technology, to be called the *Coastal Engineering Manual (CEM)*. Included in the *CEM* are the basic principles of coastal processes, methods for computing planning and design parameters, and guidance on how to develop and conduct studies in support of coastal storm damage reduction, shore protection, and navigation projects. Broader coverage of all aspects of coastal engineering are provided, including new sections on navigation and harbor design, dredging and dredged material placement, structure repair and rehabilitation, wetland and low energy shore protection, cohesive shores, risk analysis, numerical simulation, the engineering process, and other topics.

I-4-2. Structure

The *CEM* contains two major subdivisions: science-based parts and engineering-based parts. The science-based parts include "Part II - Coastal Hydrodynamics," "Part III – Coastal Sediment Processes," and "Part IV – Coastal Geology." These provide the scientific foundation on which the engineering-based parts rely.

a.   Part II. "Coastal Hydrodynamics" is organized to lead the reader from the fundamental principles of linear and other wave theories, including irregular waves and spectral analysis, to ocean wave generation and through the process of transformation as the wave approaches and reacts with the coastline. Analysis of water level variations including astronomical tides and storm surges are presented along with the hydrodynamics of coastal inlets and harbors are included in other chapters.

b.   Part III. "Coastal Sediment Processes" includes chapters on sediment properties, along shore and cross-shore transport, as well as chapters on wind transport, cohesive sediment processes and shelf transport.

c.   Part IV. "Coastal Geology" includes chapters on terminology, geomorphology, and morphodynamics.

The two engineering-based parts, Part V – "Coastal Project Planning and Design" and Part VI – "Design of Coastal Project Elements" are oriented toward a project-type approach, rather than the individual structure design approach that characterized the *SPM*. The architecture and substance of the engineering-based parts is the result of an internationally-attended workshop in February 1994. A logical systems-based approach is used for the engineering structure of the *CEM*. This mirrors the engineering process with guidance in

I-4-4. Acknowledgments

Authors of Chapter I-4, "The Coastal Engineering Manual:"

Joan Pope, U.S. Army Engineer Research and Development Center, Vicksburg, Mississippi.
John H. Lockhart, Jr., Headquarters, U.S. Army Corps of Engineers, Washington, DC, (retired).

Reviewer:

Andrew Morang, Ph.D., CHL

www.ingramcontent.com/pod-product-compliance
Lightning Source LLC
Chambersburg PA
CBHW081341190326
41458CB00018B/6071